XXI International Youth Scientific School on Coherent Optics and Optical Spectroscopy (COOS 2017)

Journal of Physics: Conference Series
Volume 1068

Kazan, Russia
17 - 19 October 2017

Editors:

Salakhov Myakzyum Khalimullovich
Samartsev Vitalii Vladimirovich
Gainutdinov Renat Khamitovich

ISBN: 978-1-5108-6941-7
ISSN: 1742-6588

Printed from e-media with permission by:

Curran Associates, Inc.
57 Morehouse Lane
Red Hook, NY 12571

CURRAN ASSOCIATES INC. proceedings.com

Some format issues inherent in the e-media version may also appear in this print version.

Printed by Curran Associates, Inc. (2018)

Institute of Physics
Dirac House, Temple Back
Bristol BS1 6BE UK

Phone: 44 1 17 929 7481
Fax: 44 1 17 920 0979

techtracking@iop.org

Additional copies of this publication are available from:

Curran Associates, Inc.
57 Morehouse Lane
Red Hook, NY 12571 USA
Phone: 845-758-0400
Fax: 845-758-2633
Email: curran@proceedings.com
Web: www.proceedings.com

Table of contents

Volume 1068

XXI International Youth Scientific School "Coherent Optics and Optical Spectroscopy"

17-19 October 2017, Kazan, Russian Federation

Accepted papers received: 23 July 2018
Published online: 17 August 2018

Preface

Papers

IOP Conf. Series: Journal of Physics: Conf. Series **1068** (2018) 011001 doi:10.1088/1742-6596/1068/1/011001

The XXI International Youth Scientific School "Coherent Optics and Optical Spectroscopy" (COOS 2017) was held in Kazan, Russia, from October 17 to October 19 on the Nikolai Lobachevsky Scientific Library of Kazan Federal University. The School follows the global tendency to comprehensive studies of matter properties and its interaction with electromagnetic fields. Since 1997 more than 100 famous scientists from USA, Germany, Ukraine, Belarus and Russia had plenary lectures presentations. This is the right place, where over 1000 young scientists had an opportunity to participate in hot discussions of the latest scientific news. Many young people have submitted interesting reports on photonics, quantum electronics, laser physics, quantum optics, traditional optical and laser spectroscopy, non-linear optics, material science and nanotechnology. Here we are publishing the full-size papers prepared from the most interesting lectures and reports selected by the Program Committee of the School. Plenary sessions were offered by the following invited speakers:

Andrei Naumov, Institute for Spectroscopy RAS, Troitsk, Moscow, Russia.

- Electron-phonon interaction in impure disordered solid media at low temperatures: microscopic nature and manifestation in the spectra of single molecules

Sergey Sazonov, National Research Center "Kurchatov Institute", Moscow, Russia.

- On the phase and group velocities of optical solitons

Renat Gainutdinov, Kazan Federal University, Kazan, Russia.

- Quantum computer science and the laws of the microworld

Anastas Bukharaev, Kazan E. K. Zavoisky Physical -Technical Institute, Kazan, Russia.

- Scanning probe microscopy of magnetic micro- and nanostructures

Sergey Kharintsev, Kazan Federal University, Kazan, Russia.

- Nonlinear plasmonics of refractory materials

Anatolii Leukhin, Mari State University, Yoshkar-Ola, Russia.

- Space and airborne radiolocation of terrestrial mobile objects in the mode of synthesizing the antenna aperture

Askhat Basharov, National Research Center "Kurchatov Institute", Moscow, Russia.

- Interaction of broadband quantum fields with resonant atoms in second-order algebraic perturbation theory
- Generation of terahertz impulse via an excited localized ensemble of identical particles with a permanent dipole moment

PROGRAM COMMITTEE

Naumov Andrey Vitalievich – chairman
Dr. Sc. (Phys.-Math.), Professor, Scientific vice-director of the Institute for Spectroscopy RAS
Gabitov Ildar Ravilievich – co-chairman
Dr. Sc. (Phys.-Math.), Director of the Skoltech Center for Photonics and Quantum Materials, Russia, professor at University of Arizona, USA
Salakhov Myakzyum Khalimulovich
Dr. Sc. (Phys.-Math.), President of the Tatarstan Academy of Science
Samartsev Vitaly Vladimirovich
Dr. Sc. (Phys.-Math.), member of the Russian Academy of Natural Science, Professor at Kazan E. K. Zavoisky Physical-Technical Institute (KPhTI)
Gainutdinov Renat Khamitovich
Dr. Sc. (Phys.-Math.), Professor at Kazan Federal University

ORGANIZING COMMITTEE
Salakhov Myakzyum Khalimullovich – rector of the School
Samartsev Vitalii Vladimirovich – vice-rector of the School
Kamalova Dina Ilevna – scientific secretary of the School

This proceedings volume contains papers reported at the conference. The Proceedings of the XXI International Youth School on Coherent Optics and Optical Spectroscopy (COOS 2017) were edited by Salakhov Myakzyum Khalimullovich, Samartsev Vitalii Vladimirovich and Gainutdinov Renat Khamitovich with the assistance of Petrova Alexandra Sergeevna as copy editor.

We are grateful to Russian Foundation for Basic Research, Tatarstan Academy of Science, Kazan E.K. Zavoisky Physical-Technical Institute for financial.

Salakhov Myakzyum Khalimullovich, Editor
Samartsev Vitalii Vladimirovich, Editor
Gainutdinov Renat Khamitovich, Editor

IOP Conf. Series: Journal of Physics: Conf. Series **1068** (2018) 011002 doi:10.1088/1742-6596/1068/1/011002

Peer review statement

All papers published in this volume of *Journal of Physics: Conference Series* have been peer reviewed through processes administered by the proceedings Editors. Reviews were conducted by expert referees to the professional and scientific standards expected of a proceedings journal published by IOP Publishing.

IOP Conf. Series: Journal of Physics: Conf. Series **1068** (2018) 012001 doi:10.1088/1742-6596/1068/1/012001

Information losses when recording the stimulated echo-hologram in gas media

E N Akhmedshina[1], L A Nefediev[1], G I Garnaeva[1]and N E Zamaliev[1]

Department of Educational Technologies in Physics, Kazan Federal University, Kazan, 420008, Russia

Email: ekanika8@gmail.com, nefediev@yandex.ru, guzka-1@yandex.ru, niyazzamaliev@gmail.com

Abstract. Considered is the impact of collisions with the change of the particles velocity in the ^{174}Yb vapor exerted on the information reproducibility in the stimulated echo hologram response, encoded in the temporal form of the second object laser pulse. It was manifested that such collisions result in the increase of the "spurious" information, serving the background for loosing information introduced into the object laser pulse in the stimulated echo hologram response.

1. Introduction
The processof echoholograms formation in gas media is qualitatively different from its analog in a solid body [1-6]. Differences are associated with the "working" particlesmovement, which causes the need of accounting collisions and random reorientations of particles which occur at formation of optical coherent responses of the media. The velocity of individual particles in gases is a vector quantity. Thus, each point of such a sample is associated with its own direction in space (the directed nature of the Doppler effect). Therefore, the amplitude and phase information containing in the object wave turns out to be "encoded" in streams of particles moving with certain velocities. In addition, phasing of individual dipole moments in gas at the time of the echo formation does not necessarily take place at each point of the sample, but thereoccurs phasing of individual velocity "packages" of particles moving with different velocities. This paper considers the impact of collisions with the velocity change of particles in gas on the reproducibility of information in the stimulated echohologram (SEH) response at the $^1S_0(6s^2)$-$^3P_1(6s6p)$ junction in ^{174}Yb vapor, encoded in the temporal form of the second object laser pulse. Obtained is the dependence of differential information entropy of the SEH response temporal form at the change in the time interval between the first and the third laser pulses due tocollisions with change in the velocity of gas particles, as well as at different ^{174}Yb vapor pressure.

2. The basic equation
For description of the echo hologram formation in a gas, it is necessary to simultaneously account the Doppler shifts in the particles radiation frequencies, the changes in their position and orientation in space, the collision of particles, as well as the degeneration of their energy levels by over some quantum number. In the case of a shock mechanism of broadening the resonance transition line the

IOP Conf. Series: Journal of Physics: Conf. Series **1068** (2018) 012001 doi:10.1088/1742-6596/1068/1/012001

equation for the single-particle density matrix with the streaming members in the laboratory coordinate system have the form [2,3]:

$$\left(\frac{\partial}{\partial t}+\frac{\partial \Delta\vec{r}}{\partial t}\nabla\right)\rho_{\gamma\alpha}^{J_1J_2}\left(1-\frac{\vec{n}}{c}\frac{\partial \Delta\vec{r}}{\partial t}\right)^{-1}=\frac{i}{\hbar}\left(H_{J_1}^0-H_{J_2}^0\right)\rho_{\gamma\alpha}^{J_1J_2}+$$

$$+\frac{i}{\hbar}\sum_{\beta}\left(V_{\gamma\beta}^{J_1J}\rho_{\beta\alpha}^{JJ_2}-V_{\beta\alpha}^{JJ_2}\rho_{\gamma\beta}^{J_1J}\right)-N\left\langle\upsilon\left(\sigma'_{\gamma\alpha}-i\sigma''_{\gamma\alpha}\right)\right\rangle\rho_{\gamma\alpha}^{J_1J_2} \quad (1)$$

where J_i-is the total moment of i-th level, indexes α, β, γ denote sublevels degenerate levels, $\Delta\vec{r}$ is the displacement of the particle due to its movement on the considered time interval, $\vec{n}$ is the direction of propagation of the laser excitation pulse, $V_{\beta\beta'}^{JJ'}$ are the matrix elements of the interaction operator of the atom with electromagnetic radiation, H^0-is Hamiltonian of the unperturbed atom, N- is the concentration of the perturbing particles, υ is the relative velocity of the atom and the perturbing particle, σ 'and σ" are cross-sections for the broadening and shifting of the corresponding spectral line, c- is the velocity of propagation of the excitation laser pulse in the medium.

Let us consider the SEH formation in gas system considering the effect of particle motion. The best mode for recording and reproducing an echo hologram in a gas is possible in case when the sample (cells) dimensions are small as compared with the distance to the object. In this case, an decomposition of the object laser pulse field over plane waves shall become possible. The electric field amplitude of the SEH response will be:

$$E\sim\sum_j\int W(\vec{v})d\vec{v}\left\langle\ddot{\vec{d}}_j\right\rangle\times\vec{n}_n\times\vec{n}_n \quad (2)$$

where $\vec{n}_n$ -are the unit vectors in the direction of plane waves propagation, $W(\vec{v})$ -the particle velocity distribution function, $t'=t-\frac{\vec{R}_0\vec{n}_n}{c}+\frac{\vec{r}_j\vec{n}_n}{c}$, $\vec{R}_0$ -the radius vector of the observation point, and $\vec{r}_j$ - the radius vector of the j-th optical center location, $\left\langle\ddot{\vec{d}}(t')\right\rangle=Sp\left(\rho\vec{d}_j(t')\right)=\vec{d}_{21}\rho_{12}^{(3)}(t')+\kappa.c.$ where ρ_{ij} - are the matrix elements of the density matrix have been obtained in [7].

The recording of an SEH in a gas is possible in case the conditions of spatial phase synchronism have the form

$$\vec{n}_{0n}-\vec{n}_1+\vec{n}_{2n}-\vec{n}_3=0 \quad (3)$$

It was shown in [8, 9] that elastic collisions with the change in the direction of the particle velocity in gas result in spectral diffusion within the inhomogeneous broadened line, which affects the temporal form of the SEH response. Let us consider the case when the information is stored in the temporal form of the second object laser pulse at the SEH recording. The pulse sequence at excitation of a stimulated echo hologram is shown in figure 1.

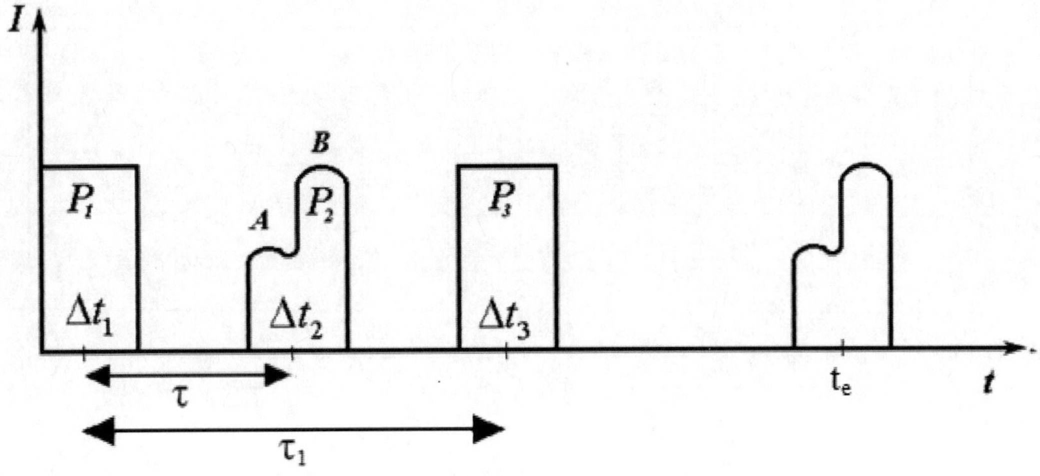

Figure 1. The sequence of pulses at SEH excitation. Δt_i - is the duration of the *i*-th resonant laser pulse, t_e - is the time of appearance of the SEH response, $\tau = t_2 - t_1$ and $\tau_1 = t_3 - t_1$ are the time intervals between the exciting pulses, t_i -moment of the *i*-th laser pulse impacts, P_1, P_2, P_3 - resonance laser pulses.

The electric field amplitude of the SEH response in case of the linear polarization of the excitating laser pulses will be determined by an expression similar to the one obtained in [8]:

$$E \sim \int_{-\infty}^{\infty} \frac{\sin(\theta_1(\Delta))\sin(\theta_2(\Delta,\delta\varphi_\tau(\Delta)))\sin(\theta_3(\Delta,\delta\varphi_{\tau_1}(\Delta)))}{\theta_1'(\Delta)\theta_2'(\Delta,\delta\varphi_\tau(\Delta))\theta_3'(\Delta,\delta\varphi_{\tau_1}(\Delta))} \times$$
$$\tilde{S}_1^*(\Delta)\tilde{S}_2(\Delta)\tilde{S}_3(\Delta)\exp\{i\Delta[t\cdot\cos(\delta\varphi_t(\Delta)) - \tau\cdot\cos(\delta\varphi_\tau(\Delta)) - \tau_1\cdot\cos(\delta\varphi_{\tau_1}(\Delta))]\}g(\Delta)d\Delta, \quad (4)$$

where

$g(\Delta) = \left(1/\Delta_D\sqrt{2\pi}\right)\exp\left\{-(\Delta/\Delta_D)^2\right\}$, $\theta_1(\Delta) = \theta_1'(\Delta)\delta t_1$, $\theta_2(\Delta,\delta\varphi_\tau(\Delta)) = \theta_2'(\Delta,\delta\varphi_\tau(\Delta))\delta t_2$, $\theta_3(\Delta,\delta\varphi_{\tau_1}(\Delta)) = \theta_3'(\Delta,\delta\varphi_{\tau_1}(\Delta))\delta t_3$ are areas of laser pulses, $\tilde{S}_1(\Delta)$, $\tilde{S}_2(\Delta)$, $\tilde{S}_3(\Delta)$ are Fourier spectra of pulses, $\delta\varphi_\tau(\Delta)$, $\delta\varphi_{\tau_1}(\Delta)$, $\delta\varphi_t(\Delta)$ is the average random increase in the angle between the velocity of the molecule and the direction of observation during corresponding time interval and the corresponding isochrome of the Doppler-broadened line, Δ_D is the width of the heterogeneously (Doppler) broadened line, τ is the time interval between the first and the second pulses, τ_1 - time interval between the 1st and 3rd pulses.

Numerical calculation of the expression (4) at information encoding in the temporal form of the object pulse for the SEH response is shown in figure 2 (a, b), at changing the temporal interval between 1 and 3 pulses.

IOP Conf. Series: Journal of Physics: Conf. Series **1068** (2018) 012001 doi:10.1088/1742-6596/1068/1/012001

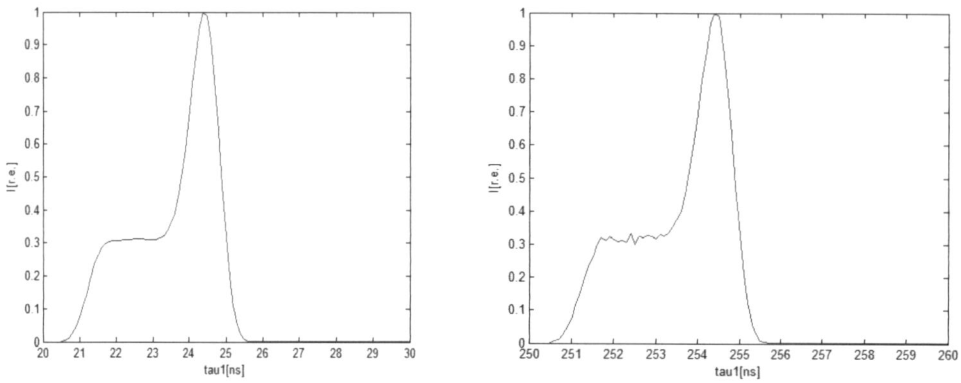

a) b)

Figure 2. Temporal form of SEH response in the ^{174}Ybvapor τ=5нс, T=800°K, P=1,3 Торр a) $\tau_1 = 20\,\text{ns}$ b) $\tau_1 = 250\,\text{ns}$.

From the analysis of the figures it follows that the information display in the SEH response encoded in the temporal form of the second object pulse deteriorates at changing the time interval between 1 and 3 pulses on the $^1S_0(6s^2)$-$^3P_1(6s6p)$ transition of the ^{174}Yb vapor. Note that an increase in pressure results in a similar effect [8, 9]. We introduce differential information entropy in order to describe the number of reproduced and recorded information which stored in the temporal form of the object laser pulse.

3. Classical differential information entropy of object laser pulse and seh response

The differential informational entropy of the temporal form of the object laser pulse is defined as:

$$J'_c = J_c - J_{c0} \tag{5}$$

where

$$J_c = -\int_0^{\delta t} p(t)\log_2 p(t)dt - \lim_{\Delta t \to 0} \log_2 \Delta t, \tag{6}$$

$$J_{c0} = -\int_0^{\delta t'} p_o(t)\log_2 p_o(t)dt - \lim_{\Delta t \to 0} \log_2 \Delta t, \tag{7}$$

δt - pulse duration, $\delta t'$ – normalization pulse duration, *p(t)* - the probability density of the distribution of amplitude (or intensity) of the electric field of pulse, $p_0(t)$ is the probability density of the distribution of amplitude (or intensity) of the electric field of the normalizing pulse, chosen against the information encoding type in the object pulse.

Differential information entropy of the temporal form of the SEH response is determined similarly to (5) with allowance for (6), where *p(t)* is the probability density of the distribution of amplitude of the electric field of the response.

A numerical calculation of the expression (5) with allowance for (4, 6-7) is shown in figure 3(a) with a change in the temporal interval between the 1st and 3-rd pulses, and in figure 3(b) underthe change of ^{174}Yb vapor pressure.

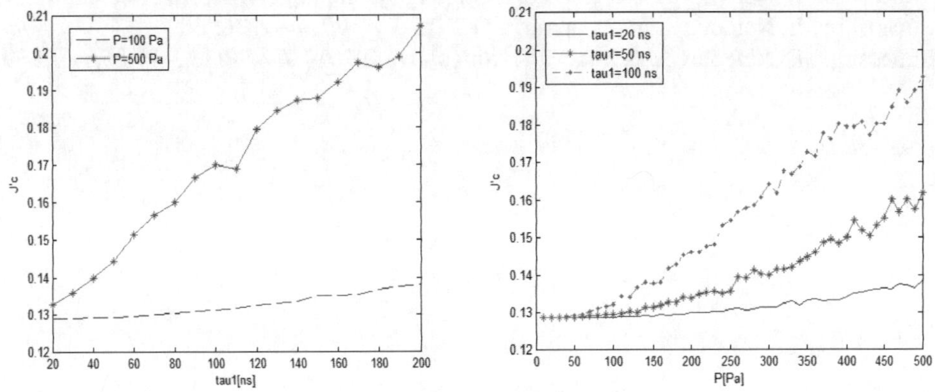

a) b)

Figure 3. The change of the differential information entropy of the stimulated echo hologram response a) the dependence on the time interval between the 1- st and 3- rd laser pulses due to collisions with a change in the velocity of the gas particles; b) the dependence on the vapor pressure value of ^{174}Yb

The change of value of differential information entropy for the stimulated echo hologram response a) dependence of value of differential information entropy on the time interval between the 1- st and 3-rd of exciting laser pulses due to collisions with a change in the velocity of the gas particles; b) dependence of value of differential information entropy on the vapor pressure value of ^{174}Yb.

Analysis of figure 3 shows that collisions with changes in the velocity of gas particles lead to the increase in "spurious" information, serving the background for losing in the SEH response of information introduced into the object laser pulse.

4. Conclusions

In the considered numerical experiments, it was found that collisions with a change in the velocity of gas particles have an impact on the temporal form of the SEH response, which results in the destruction of the recorded information in gas media. In addition, such collisions result in an increase in "spurious" information, serving the background for information losing in the SEH response that which must be taken into account when creating optical memory devices and echo processors.

Acknowledgments
The work is performed according to the Russian Government Program of Competitive Growth of Kazan Federal University

References

[1] Zuykov V,Nefed'evL, SamartsevV1982 *Applied questions of holography* (L.:LINF)p 179
[2] Nefed'evL 1985 *Opt. and Spectrosc.***59** 841

[3] Nefed'evL 1986 *J. Appl. Spectr.***44** 664
[4] Nefed'evL 1986 *Opt. and Spectrosc.***61** 387
[5] Evseev I, Rubtsova N and Samartsev V 2009 Photon echo and phase memory in gases (Kazan: Kazan State University Press) p 490
[6] Rubtsova N, Ishchenko V, Khvorostov E, Kochubei S, Khvorostov E and Yevseyev I 2004 *J.Phys. Rev.* **70** 023403
[7] Akhmedshina E, Nefediev L and Garnaeva G 2016 *Opt. and Spectrosc.***121** 405
[8] Akhmedshina E, Nefediev L and Garnaeva G 2015 *J. Appl. Spectr.***82** 669
[9] Akhmedshina E, Nefediev L, Garnaeva G, Shigapova E 2017 *J. of Phys.:Conf.Series.***859** 1

Generation of terahertz impulse via an excited localized ensemble of identical particles with a permanent dipole moment

A M Basharov

Moscow Institute of Physics and Technology, 141700 Dolgoprudny, Russia
National Research Centre «Kurchatov Institute», 123182 Moscow, Russia

E-mail: basharov@gmail.com; http://basharov.me

Abstract. The paper states a simple and analytically solvable model for the generation of low-frequency emission via a localized excited ensemble of identical two-level quantum particles with a permanent dipole moment in conditions of collective spontaneous emission of the atomic ensemble.

1. Introduction

Electromagnetic processes in media with a permanent dipole moment (PDM), for example, in ensembles of quantum dots, are believed to be attractive both from the point of view of practical applications (generation of low-frequency terahertz emission) and from the theoretical point of view. Nevertheless, many theoretical papers are confined to numerical modeling for generation of low-frequency emission. Few exceptions are the papers which consider various questions of integrability of equations, see, for example, [1-5], or processes in optically thin media [6] and microcavities, for example [7]. These papers consider dynamics of quantum systems which consist of isolated PDM-based quantum particles, where the dipole-dipole interaction is neglected either due to sparseness of the system or any other reason. In particular, such assumptions can be made so as not to account possible phase transitions associated with the spontaneous occurrence of macroscopic polarization. Then dynamics of PDM-systems is reduced to the ordinary *N*-level optical models where certain quantum energy states are parity-free and are characterized as having PDM. Unlike *N*-level systems, an entire hierarchy of multi-quantum transitions is possible in PDM-systems for all possible resonances, with the time dynamics of a quantum system being characterized by low-frequency emission processes along with high-frequency (optical) components. The origin of low-frequency components can be easily comprehended on the basis of algebraic perturbation theory - according to such an approach, dynamics of the population of energy levels is characterized by slow time changes in comparison with the period of rapid oscillations of values determined by optical quantum transitions. Then, since particles have PDM, there arises an ordinary dipole generation of low-frequency emission.

It is to be noted that an adequate apparatus for the theory of optical resonance is the algebraic perturbation theory [8-10], which makes it possible both to obtain effective Hamiltonians of various optical processes and to describe all possible (time-frequency) hierarchies of polarizations of a quantum system. Under the given conditions the generation of terahertz emission is a side effect of superradiance of excited identical two-level PDM-based atoms in the ground and/or excited energy states. The ensemble of two-level atoms is localized in a region with dimensions much smaller than the characteristic superradiance wavelength. Therefore, the dipole-dipole interaction does not manifest itself as the master equations for the populations of the energy levels under consideration. The suppression of superradiance, which becomes apparent in ensembles of a sufficient number of atoms, is also taken into account [11], which can be regarded as an additional control factor for the low-frequency emission.

2. Model definition

Suppose there are N_a fixed identical atoms in the subspace with dimensions much smaller than the characteristic wavelength, with some levels of PDM-based atoms. The Hamiltonian for such a system in the quantized electromagnetic field is given by a typical expression that includes the Hamiltonian of an isolated atoms H^A, the Hamiltonian of the quantized electromagnetic field H^F, and the operator of their interaction H^{Int}:

$$H^{Ini} = H^A + H^F + H^{Int},\ H^A = \sum_{i,j} E_j \,|\, E_j >^{(i)} < E_j \,|^{(i)},\ H^F = \sum_{\vec{q}} \hbar\omega_{\vec{q}} b^+_{\vec{q}} b_{\vec{q}}, \tag{1}$$

$$H^{Int} = \sum_{\vec{q}} \Gamma_{\vec{q}} (b^+_{\vec{q}} + b_{\vec{q}}) \sum_{i,kj} d_{kj} \,|\, E_k >^{(i)} < E_j \,|^{(i)},\ \sum_j |\, E_j >^{(i)} < E_j \,|^{(i)} = 1^{(i)},\ < E_j \,|^{(i)}\, E_k >^{(i)} = \delta_{jk}.$$

Here $|\, E_j >$ is the atomic quantum state of the energy E_j, $d_{kj} =< E_k \,|\, d \,|\, E_j >$, $d = \sum_{kj} d_{kj} \,|\, E_k >< E_j \,|$ is the dipole moment operator of an atom. The superscript of the atomic state vectors denotes the state space of the *i*-th atom and the summation over *i* is the summation over all atoms of the ensemble. Operators of creation $b^+_{\vec{q}}$ and annihilation $b_{\vec{q}}$ of photons with a wave vector $\vec{q}$ satisfy the typical commutation relation $[b_{\vec{q}}, b^+_{\vec{q}'}] = \delta_{\vec{q}\vec{q}'}$, the dispersion relationship is given by the condition $\omega_{\vec{q}} = qc$. The coupling parameter of the electromagnetic field and atoms in three-dimensional space is $\Gamma_{\vec{q}} = (2\pi\hbar qc / \ell^3)^{1/2}$, where ℓ^3 is the quantization volume.

We will speak of two-level atoms, taking into account all the energy atomic states: a pair of levels is allocated by initial conditions - only the given atomic level $|\, E_2 >$ is assumed to be populated at the initial instant of time, which is coupled by the frequency transition $\omega_0 = (E_2 - E_1)/\hbar$ with the low-lying ground level. For the sake of simplicity, let us suppose that there are no other single-quantum transitions from the excited level to the low-lying ones. In contrast to particles whose quantum states are characterized by parity, two-quantum transitions are possible in PDM-based media, along with single-quantum transitions from the excited level $|\, E_2 >$ to the ground level $|\, E_1 >$, since there is an entire frequency spectrum in the vacuum electromagnetic field of ordinary three-dimensional space. However, two-quantum processes are second-order processes in the coupling constant of a quantum particle with an electromagnetic field and let us assume that there is no reason for a serious increase in the significance of such two-quantum processes in comparison with single-quantum processes in the system under consideration, as is the case for the same parity of states $|\, E_1 >$ and $|\, E_2 >$, or in the case of processes in media with spectral singularities [12].

We will construct an effective Hamiltonian and obtain the master equation for the atoms. In case of open systems the major technique for constructing an effective Hamiltonian is the algebraic perturbation theory [8-10].

In the extended space of states of the atomic system and the quantized electromagnetic field, the initial Hamiltonian H^{Ini} determines the Schrödinger equation for the state vector: $i\frac{d}{dt} |\, \Psi^{A+F} >= H^{Ini} \,|\, \Psi^{A+F} >$. According to the algebraic perturbation theory and methods of the effective Hamiltonian, assumptions about the nature of the initial state $|\, \Psi^{A+F} >$ at $t = 0$ are made after constructing the effective Hamiltonian of the problem.

3. The effective Hamiltonian, the master equation and low-frequency polarization of atoms

Following the general scheme [8-10], we transform the state vector $|\Psi^{A+F}>$:

$$|\tilde{\Psi}^{A+F}>=U|\Psi^{A+F}>,\ U=e^{-iS},\ S^{+}=S. \tag{2}$$

The transformed vector will satisfy the Schrödinger equation

$$i\hbar\tfrac{d}{dt}|\tilde{\Psi}^{A+F}>=\tilde{H}|\tilde{\Psi}^{A+F}>$$

with the transformed Hamiltonian

$$\tilde{H}=UH^{Int}U^{+}-i\hbar U\tfrac{d}{dt}U^{+}.$$

Expanding $\tilde{H}$ and S in series over the interaction with the vacuum field

$$S=S^{(1)}+S^{(2)}+\ldots,\quad \tilde{H}=\tilde{H}^{(0)}+\tilde{H}^{(1)}+\tilde{H}^{(2)}+\ldots,$$

we obtain (taking into account the Baker-Hausdorff formula [9])

$$\tilde{H}^{(0)}=H^{A}+H^{F},\ \tilde{H}^{(1)}=H^{Int}-i[S^{(1)},\tilde{H}^{(0)}]+\hbar\tfrac{d}{dt}S^{(1)},$$

$$\tilde{H}^{(2)}=-\tfrac{i}{2}[S^{(1)},H^{Int}]-\tfrac{i}{2}[S^{(1)},\tilde{H}^{(1)}]-i[S^{(2)},\tilde{H}^{(0)}]+\hbar\tfrac{d}{dt}S^{(2)}. \tag{3}$$

The major idea for finding the terms $S^{(i)}$ and the effective Hamiltonian $H^{Eff}=\tilde{H}^{(0)}+\tilde{H}^{(1)}+\tilde{H}^{(2)}$ is the absence of rapidly varying terms in $\tilde{H}^{(i)}$ in the interaction representation. The author's approach [8-10] differs from the other approaches to constructing the effective Hamiltonian of the theory of open quantum systems [13, 14] and leads to the notion that the algebraic perturbation theory has become self-consistent and efficient to describe both open quantum systems and quantum and nonlinear optical phenomena. Additionally, for open systems the differences are related to the Markov conditions that can be imposed on the initial non-transformed state vectors or the transformed state vectors. The differences arising here are vividly illustrated by the model problem considered in [15].

The conventional [8-11] calculations give rise to the terms in the second order of the effective Hamiltonian which are diagonal according to atomic variables and do not depend on photon variables (Lamb shifts). They can be easily included into $\tilde{H}^{(0)}$. Then projection operators associated with the non-resonant energy levels can be omitted in the second-order term $\tilde{H}^{(2)}$.

A term describing the dipole-dipole interaction of atoms tends to appear [11] in the second order of algebraic perturbation theory in case of an atomic interaction operator with a quantized electromagnetic field. This dipole-dipole interaction operator is neglected, since it does not exert a serious impact on the master equation for resonance level populations which is to be treated below. This leads to the following expressions (the tilde is also omitted):

$$H^{(0)}=\hbar\omega_{21}'R_3,\ \omega_{21}'=(E_2'-E_1')/\hbar,\ H^{(1)}=\sum_{\omega}\Gamma_\omega b_\omega^{+}d_{12}R_{-}+\sum_{\omega}\Gamma_\omega b_\omega d_{21}R_{+},$$

$$H^{(2)}=\tfrac{N_a}{2}\sum_{\omega}\Gamma_\omega b_\omega^{+}\sum_{\omega'}\Gamma_\omega b_{\omega'}(\Pi(\omega)+\Pi(\omega')), \tag{4}$$

$$E_1'=E_1+\sum_{\omega}\Gamma_\omega^2\sum_{j}\frac{|d_{1j}|^2}{\hbar(\omega_{1j}-\omega)}|E_1><E_1|,\ E_2'=E_2+\sum_{\omega}\Gamma_\omega^2\sum_{j}\frac{|d_{2j}|^2}{\hbar(\omega_{2j}-\omega)}|E_2><E_2|.$$

During the calculations, unified parameters of the theory of optical resonance processes [8-10] are emerging regardless of the presence of classical and/or quantized fields, the presence or absence of PDM

$$\Pi_k(\omega)=\sum_{j}{}'\frac{|d_{kj}|^2}{\hbar}\left(\frac{1}{\omega_{kj}+\omega}+\frac{1}{\omega_{kj}-\omega}\right). \tag{5}$$

The prime at the summation sign means that the summation eliminates the resonance terms $\omega_{jk}\mp\omega\approx 0$. In addition, the definition is used $\omega_{km}=(E_k-E_m)/\hbar$.

The formula for the term of the effective Hamiltonian $H^{(2)}$ describing the Stark interaction suggests that the condition $\Pi_1(\omega) \approx \Pi_2(\omega) \equiv \Pi(\omega)$ is satisfied for the resonance levels $|E_1>$ and $|E_2>$.

The expression for $\Pi_k(\omega)$ contains no restrictions [8-11] for parity of the energy levels involved. Nevertheless, it is obvious that if only a pair of levels $|E_1>$ and $|E_2>$ were considered in the initial problem statement, then the presence of PDM would not have any effect on the system dynamics in the optically thin medium. The presence of nonresonant levels different from the considered ones leads to a change in the parameter $\Pi_k(\omega)$, triggers the dependence of this parameter on PDM and as result it becomes possible to discuss the PDM-based system dynamics for various values $\Pi_k(\omega)$.

The effective Hamiltonian $H^{Eff} = \tilde{H}^{(0)} + \tilde{H}^{(1)} + \tilde{H}^{(2)}$ is represented by atomic operators

$$R_3 = \tfrac{1}{2}\sum_i (|E_2>^{(i)}<E_2|^{(i)} - |E_1>^{(i)}<E_1|^{(i)}),\ R_- = \sum_i |E_1>^{(i)}<E_2|^{(i)},\ R_+ = \sum_i |E_2>^{(i)}<E_1|^{(i)},$$

satisfying commutation conditions $[R_3, R_\pm] = \pm R_\pm$, $[R_+, R_-] = 2R_3$.

The polarization of the medium P is expressed in terms of the transformed density matrix, $\rho = \mathrm{Tr}_F |\tilde{\Psi}^{A+F}><\tilde{\Psi}^{A+F}|$ where the trace is taken over the states of the electromagnetic field. The general formula for polarization $P = N_a \mathrm{Tr}\rho D$ is determined by the effective dipole moment operator

$$D = e^{-iS} d e^{iS} = d - i[S,d] - \tfrac{1}{2}[S,[S,d]] + ... = \sum_{kj} D_{kj} |E_k><E_j|.$$

The low-frequency (LF) polarization component emerging in PDM-based media is of interest for our purposes. In this case, it can be written in the first non-vanishing order of the algebraic perturbation theory in its simple form

$$P^{LF} = N_a(d_{11}\rho_{11} + d_{22}\rho_{22}). \qquad (6)$$

To calculate polarization P^{LF} and determine parameters of the generated low-frequency emission it is necessary to obtain the master equation for the atomic density matrix. In case of absence of PDM, the formulated problem was solved in [11]. The results of algebraic perturbation theory make it possible to use the results of [11] under the conditions defined above and to immediately write down the master equation for the atomic density matrix according to the approximations described above.

We will consider the states of the atomic ensemble, symmetric in permutations of atoms. In that case they can be conveniently described by eigenvectors $|m>$ of the operator R_3: $R_3|m> = m|m>$, $-r \le m \le r$. Here r is the parameter characterizing the irreducible representation of the angular momentum algebra $su(2)$ for describing symmetric states of an ensemble of $2r = N_a$ identical atoms [11,16]. The state of an ensemble of fully excited quantum particles is given by a vector $|r>$, and the state of an ensemble, whose particles are in the ground state, is given by a vector $|-r>$.

It is not difficult to write down master equations satisfying the effective Hamiltonian presented above in the Markov approximation [11] in the form:

$$\frac{d\rho_{mm}}{d\tau} = 2\chi^2(\eta_0 N_a)^{-2}[1-\cos(\eta_0 N_a)](g_{m+1\,m}\rho_{m+1m+1} - g_{m\,m-1}\rho_{mm}),\ g_{m\,m-1} = (r+m)(r-m+1). \qquad (7)$$

Here the dimensionless parameters are involved

$$\tau = \omega_{21} t,\ \chi = \sqrt{2}\omega_{21} d_{12} \mu^{-1} c^{-3/2} \hbar^{-1/2},\ \eta_0 = 4\omega_{21}^3 \mu^{-2} c^{-3} \Pi(\omega_{21}).$$

Parameter $\mu \sim 1$ was involved to describe various models for a quantized electromagnetic field [11].

It can be seen that in neglecting the two-quantum decay of the excited level $|E_2>$ the main difference of the dynamics of the PDM-based system under study lies both in the magnitude η_0 and in inducing the low-frequency polarization (6), which takes the following form in terms of the introduced representation (henceforward everything is again expressed in dimensional values):

$$P^{LF} = N_a \sum_{m=-r}^{r} [d_{22}(r+m) + d_{11}(r-m)]\rho_{mm} = p^{LF}(t) + \text{const}, \; p^{LF}(t) = N_a (d_{22} - d_{11}) \sum_{m=-r}^{r} m\rho_{mm} \, . \quad (8)$$

4. Impulse of low-frequency emission under superradiance

At the initial instant of time $t = 0$ all quantum particles of the ensemble are assumed to be excited, so that only the following one is nonzero $\rho_{rr}(0) = 1$ among the matrix elements of the density matrix. These initial conditions satisfy the Dicke ordinary superradiance [16], however, in contrast to the Dicke superradiance in Eqs. (7), the effect of collective-relaxation suppression by the Stark interaction, which is clearly manifested in localized atomic ensembles [11], is taken into account. In addition, equations (7) and (8) account for the presence of PDM-based atoms.

Let us consider a low-frequency component of the dipole emission of an atomic ensemble. For that purpose we will use the classical formula for the total intensity of dipole emission [17]. Using the dimensional values we have:

$$I^{LF} = \tfrac{2}{3} c^{-3} (d^2 P^{LF}(t)/dt^2)^2 = \tfrac{2}{3} c^{-3} N_a^2 (d_{22} - d_{11})^2 (\tfrac{d^2}{dt^2} \sum_{m=-r}^{r} m\rho_{mm})^2 .$$

The value $\frac{d}{dt} \sum_{m=-r}^{r} m\rho_{mm}$ that determines the intensity of low-frequency emission is proportional to the rate of change in the average energy of the atomic system and, as a consequence, the intensity $I^{SR}(t)$ of high-frequency superradiance:

$$\tfrac{d}{dt} \sum_{m=-r}^{r} m\rho_{mm} = -(\hbar\omega_{21})^{-1} I^{SR}(t) \, , \; I^{SR}(t) = \hbar\omega_{21} \tfrac{d}{dt} n \, .$$

As a result, we obtain the relation between the required intensity of the low-frequency emission and the intensity of the superradiance of the atomic ensemble in the approximations under consideration:

$$I^{LF} = \tfrac{2}{3} c^{-3} (d^2 P^{LF}(t)/dt^2)^2 = \tfrac{2}{3} c^{-3} \hbar^{-2} \omega_{21}^{-2} N_a^2 (d_{22} - d_{11})^2 (dI^{SR}(t)/dt)^2 .$$

To obtain the expression for the average number of n photons of the frequency ω_{21} radiated by the time moment t, we use the estimate of the mean emission time $\bar{t}_n$ of these n photons as

$$\bar{t}_n = \frac{(\eta_0 N_a)^2}{2\chi^2 \omega_{21} [1 - \cos(\eta_0 N_a)]} \sum_{n=r/2-n+1}^{r/2} g_{m,m-1}^{-1} \, .$$

Under the conditions of suppression of collective relaxation processes, the intensity of superradiance at $N_a >> 1$ can be estimated as

$$I^{SR}(t) \approx \bar{\gamma} \hbar\omega_{21} \tfrac{1}{4} N_a^2 \text{sech}^2 [\bar{\gamma} N_a (t - t_D)/2], \; \bar{\gamma} = 2\chi^2 \beta \omega_{21} (\eta_0 N_a)^{-2} [1 - \cos(\eta_0 N_a)], \; t_D = (\bar{\gamma} N_a)^{-1} \ln \bar{\gamma} N_a ,$$

where β is the dimensionless geometry factor.

Eventually, the intensity of low-frequency emission takes the following form

$$I^{LF} = \frac{2(d_{22} - d_{11})^2}{3c^3} \omega_{21}^4 N_a^8 \chi^8 \beta^4 \frac{(1 - \cos(\eta_0 N_a))^4}{(\eta_0 N_a)^8} \frac{\text{sh}^2 [\bar{\gamma} N_a (t - t_D)/2]}{\text{ch}^6 [\bar{\gamma} N_a (t - t_D)/2]} \, . \quad (9)$$

Compared with the intensity of ordinary superradiance, low-frequency generation depends heavily on the number of atoms in the ensemble ($I^{LF} \sim N_a^8$) and appears to be of a two-humped form of a zero value corresponding to the maximum intensity of the superradiance pulse. Time interval T between the maxima of low-frequency emission can be estimated as $T = 2(\bar{\gamma} N_a)^{-1} \ln(2 + \sqrt{3})$ and as the number of atoms increases, the distance between the peaks of low-frequency emission as well as their duration drop to zero value. In case of a significant number of atoms $N_a \sim \eta_0^{-1}$ the emission intensity begins to decrease, and at $N_a = 2\pi\eta_0^{-1}$ vanishes similar to the high-frequency superradiance [11].

Comparison of the derived formulae with the other cases of low-frequency generation emission under single-quantum resonance conditions [6,7] shows that the intensity of low-frequency emission depends on the difference of PDM values in the ground and excited states $d_{22}-d_{11}$.

5. Conclusion

The algebraic perturbation theory for optical processes provides a description of optical processes under resonant conditions, which causes the slow dependence of the populations of quantum levels on time, in comparison with the rapid time changes of optical properties within a quantum system. Then the quantum system always has a low-frequency component of polarization in PDM-based systems. Thus, the presented model analysis of the generation of low-frequency emission allows us to make a clear distinction between the low-frequency emission as a side action of optical effects in PDM-based media and the generation of low-frequency emission caused by the inherent property of the generated medium. And although the presence of PDM is often an integral property of the medium, the explanation of generation of terahertz emission in this case reduces frequently to the derivation of this very permanent dipole moment. Meanwhile, the very fact of generation of terahertz emission in optical processes in optically thin media with a permanent dipole moment does not require any new considerations other than the typical application of algebraic perturbation theory (the basis of the theory of optical resonance [10-12]) and the classical theory of dipole emission [17] (or certain approximations, for example, the approximation of unidirectional waves [18]).

Acknowledgments

This work is supported by the Russian Foundation for Basic Research (grant No. 16-02-00453a).

References

[1] Agrotis M et al 2000 Physica D **138** 134
[2] Maimistov A I and Caputo J-G 2003 Optics and Spectroscopy **94** 245
[3] Zabolotskiĭ A A 2008 JETP **106** 846
[4] Sazonov S V 2016 Optics Communications **380** 480
[5] Sazonov S V and Ustinov N V 2017 JETP **124** 213
[6] Basharov A M 2016 JETP Lett. **103** 15
[7] Chestnov I Yu et al 2016 JETP Lett. **104** 169
[8] Basharov A M, Maimistov A I and Manykin E A 1983 Sov.Phys. JETP **57** 282
[9] Maimistov A I and Basharov A M 1999 *Nonlinear optical waves* (Dordrecht: Kluwer Academic)
[10] Basharov A M 2012 JETP **115** 371
[11] Basharov A M 2011 Phys.Rev. A **84** 013801
[12] Basharov A M 1999 JETP Lett. **70** 438
[13] Accardi L, Lu Y G and Volovich I 2002 *Quantum theory and its stochastic limit* (Berlin, Heidelberg, GmbH: Springer-Verlag)
[14] Breuer H-P and Petruccione F 2002 *Theory of open quantum systems* (Oxford: OUP)
[15] Basharov A M 2017 J Phys. CS **859** 012003
[16] Benedict M G et al 1996 *Super-radiance* (Bristol and Philadelphia: IOP)
[17] Landau L D and Lifshitz E M 1994 *The classical theory of fields* (Butterworth-Heinemann)
[18] Belenov E M and Nazarkin A V 1990 JETP Lett. **51** 288

Interaction of broadband quantum fields with resonant atoms in second-order algebraic perturbation theory

A M Basharov

Moscow Institute of Physics and Technology, 141700 Dolgoprudny, Russia
National Research Centre «Kurchatov Institute», 123182 Moscow, Russia

E-mail: basharov@gmail.com; http://basharov.me

Abstract. The second-order algebraic perturbation theory provides a correct and conventional description of the interaction of a single-mode quantized field and a broadband field with resonant particles, and in case of interacting two different broadband fields in the Markov approximation the generalization of the Hudson-Parthasarathy algebra is required to derive the Schrödinger equation.

1. Introduction

The development of media preparation technologies, artificial emitters and quantum particles has led to an experimental study of numerous interactions of various electromagnetic fields with quantum particles in various resonance conditions. In nonlinear optics [1], the interaction of classical fields with N-level (multilevel) quantum particles is effectively considered by the algebraic perturbation theory [2-4], which is an algebraic statement of the Bogolyubov-Krylov-Mitropolsky averaging method [5,6]. If one describes the effect of quantized broadband fields along with classical fields, then within the algebraic perturbation theory a broadband field is represented as a collection of independent sources [7]. Such a representation is important in view of the fact that the Markov approximation is used to solve the problems dealing with the effect of broadband fields, with the latter acting as noise sources and the targeted objects as open quantum systems. The broadband field as a set of independent noise sources was represented by Lax [8]. However, such a natural partitioning is not observed in other approaches [9, 10] to the problem of the resonant interaction of broadband fields with N-level quantum particles.

The description of processes of broadband field resonance interaction in the Markov approximation leads to either a restatement of the Schrödinger equation in terms of stochastic differential equations (SDEs) [11] or to a transition to master equations, with the master equations directly follow from SDE.

So far the interaction of broadband fields in the Markov approximation has been considered in terms of Wiener-type SDEs. These equations were derived when using the resonance approximation (the first-order algebraic perturbation theory) and the algebra of quantum random processes introduced by Gardiner and Colet [11]. In the case of the action of two different broadband fields, according to the first- order algebraic perturbation theory, rescattering quanta from one broadband field to the other is described as a real absorption process of one quantum with the atomic quantum transition and the emission of another quantum with an inverse atomic quantum transition. However, it is possible to rescatter the quantum even without atomic quantum transitions. Such processes are described by the second-order algebraic perturbation theory.

The second order of algebraic perturbation theory occurred [7] to describe the resonant interactions of a quantum system with coherent classical and broadband quantized fields. Models of a Raman spontaneous transition between two energy levels of a quantum particle with the absorption of a quantum of a coherent field and the emission of a quantum into a broadband quantized field were constructed [7] on the basis of Wiener-type SDEs. However, in the second-order algebraic perturbation theory, in describing the effect of only one broadband resonant quantized field there are terms occurring in the effective Hamiltonian, which can not be described by means of Wiener-type quantum random processes and which, can not be neglected in spite of their smallness. In [12] these terms are represented by a quantum counting process, with SDE being derived in terms of the algebra of quantum random processes introduced by Hudson and Parthasarathy [13]. It was suggested that such SDEs should be called non-Wiener-type ones as in [12].

This paper considers the resonant interaction of two different quantized broadband fields of zero photon density with an ensemble of identical multilevel quantum particles. The effective Hamiltonian has been determined up to the second order of the algebraic perturbation theory. It has also been shown that for the correct statement of Schrödinger equations in the Markov approximation in terms of SDEs, the Hudson-Parthasarathy algebra is no longer sufficient to describe all interactions occurring in the processes. The Hudson-Parthasarathy algebra is generalized to determine a SDE describing a system of an ensemble of identical atoms and broadband fields interacting resonantly with it.

2. Second-order algebraic perturbation theory

Resonant interaction of two broadband electromagnetic fields with a localized ensemble of identical particles will be considered in an approximation in which quantum transitions between energy levels of particles can be realized only between the allocated pair of energy levels, say, $|E_1>$ and $|E_2>$. This assumption would sound reasonable if the broadband fields under consideration had a zero photon density, the atoms occupied only the indicated pair of energy levels, with the lower of which being the ground level, and there being no other energy level between these levels. All the rest levels $|E_k>$, $k \neq 1,2$ could be regarded as being non-resonant. This fact does not mean that the nonresonance levels can be neglected - they determine the parameters of the effective Hamiltonian of the system in the second-order algebraic perturbation theory. The energy levels $|E_1>$ and $|E_2>$ are assumed to have parity, so that the transition can be considered as optically allowed.

Quantum fields are considered as being narrowly directed in order to be treated as independent and to be described by two pairs of creation and annihilation operators $b_j^+(\omega)$ and $b_j(\omega)$. The conditions which make possible to introduce operators $b_j^+(\omega)$ and $b_j(\omega)$ satisfying the typical commutation relations $[b_j(\omega), b_{j'}^+(\omega')] = \delta_{jj'}\delta(\omega-\omega')$ are considered in [12,14]. The operators $V_j(t)$ of atomic electrodipole interaction with such fields are taken in the form (in the Dirac picture):

$$V_j(t) = -\int d\omega (b_j(\omega)e^{-i\omega t} + b_j^+(\omega)e^{i\omega t})\Gamma_j(\omega)\sum_{s,km} d_{km}e^{i\Omega_{km}t}\,|E_k>^{(s)}<E_m|^{(s)}. \quad (1)$$

Here, the index $s = 1,\ldots,N_a$ marks the atoms of an ensemble localized at the origin of coordinates; the sum is calculated over one-atom interaction operators with a quantized electromagnetic field. The parameters of that interaction are determined by the matrix elements d_{km} of the dipole moment of the atom and the parameter $\Gamma_j(\omega)$ of the j-th quantized electromagnetic field. Formula (1) provides the summation over all the quantum states of an individual atom $|E_k>$, $\Omega_{km} = (E_k - E_m)/\hbar$.

One of the fundamental physical assumptions of the theory of open quantum systems is the agreement on the objects for which the Markov approximation is stated. Quantum theory is based on the Schrodinger equation for the state vector $|\Psi^{A+F}>$ of an atomic ensemble (A) and the quantized fields (F) interacting with it. In the Dirac picture we have:

$$i\hbar\frac{\partial|\Psi^{A+F}(t)>}{\partial t}=(V_1(t)+V_2(t))|\Psi^{A+F}(t)>. \quad (2)$$

However, the quantum theory has a unitary symmetry: an initial state vector $|\Psi^{Ini}(t)>=|\Psi^{A+F}(t)>$ and the initial Hamiltonian $H^{Ini}(t)=V_1(t)+V_2(t)$ can be transformed without changing the values of the observed values:

$$|\tilde{\Psi}^{A+F}(t)>=U(t)|\Psi^{A+F}(t)>,\ i\hbar\frac{\partial|\tilde{\Psi}^{A+F}(t)>}{\partial t}=\tilde{H}(t)|\tilde{\Psi}^{A+F}(t)>, \quad (3)$$

$$\tilde{H}(t)=U(t)(V_1(t)+V_2(t))U^+(t)-i\hbar U(t)\tfrac{\partial}{\partial t}U^+(t).$$

One can even perform continuous unitary transformations for diagonalizing the Hamiltonian of the entire system [15-17], but to state the Markov approximation it is necessary to indicate the state vector and the effective Hamiltonian. Paper [18] provides calculation of a simple model for the interaction of a broadband field with a two-level system in the Markov approximation and shows that the result depends on whether the Markov approximation is applied to the original or transformed values.

It is reasonable to assume that the Markov approximation is applied to an effective Hamiltonian that is adequate to the situation in question. The initial Hamiltonian has all possible transitions in an atom, while the Schrödinger equation itself and the initial Hamiltonian contain rapidly time-varying terms. The rapidly varying terms are still effectively averaged over the system dynamics, so that an adequate Hamiltonian should contain only slowly varying terms (in the Dirac picture). This is the fact that distinguishes the effective Hamiltonian in the resonance approximation and can be used as the basis for the determination of the terms of the effective Hamiltonian. In the first-order algebraic perturbation theory, the Markov approximation applied in calculations gives rise to the results which are consistent with the experimental data under suitable conditions [11, 18].

In order to derive the effective Hamiltonian of the problem, we expand the transformed Hamiltonian $\tilde{H}(t)$ and the generator $S(t)$ of transformation $U(t)=\exp(-iS(t))$ in a series in the interaction parameters with broadband fields

$$S(t)=S^{(1,0)}(t)+S^{(0,1)}(t)+\ldots,\quad \tilde{H}(t)=\tilde{H}^{(1,0)}(t)+\tilde{H}^{(0,1)}(t)+\tilde{H}^{(1,1)}(t)+\tilde{H}^{(2,0)}(t)+\tilde{H}^{(0,2)}(t)+\ldots,$$

then we obtain (taking into account the Baker-Hausdorff formula [3])

$$\tilde{H}^{(1,0)}(t)=V_1(t)+\hbar\frac{\partial S^{(1,0)}(t)}{\partial t},\quad \tilde{H}^{(0,1)}(t)=V_2(t)+\hbar\frac{\partial S^{(0,1)}(t)}{\partial t},$$

$$\tilde{H}^{(2,0)}(t)=-\tfrac{i}{2}[S^{(1,0)}(t),V_1(t)]-\tfrac{i}{2}[S^{(1,0)}(t),\tilde{H}^{(1,0)}(t)]+\hbar\frac{\partial S^{(2,0)}(t)}{\partial t},$$

$$\tilde{H}^{(0,2)}(t)=-\tfrac{i}{2}[Q^{(0,1)}(t),V_2(t)]-\tfrac{i}{2}[Q^{(0,1)}(t),\tilde{H}^{(0,1)}(t)]+\hbar\frac{\partial S^{(0,2)}(t)}{\partial t},$$

$$\tilde{H}^{(1,1)}(t)=-\tfrac{i}{2}([S^{(1,0)}(t),V_2(t)]+[S^{(1,0)}(t),\tilde{H}^{(0,1)}(t)]+[Q^{(0,1)}(t),V_1(t)]+[S^{(0,1)}(t),\tilde{H}^{(1,0)}(t)])+\hbar\tfrac{\partial}{\partial t}S^{(1,1)}(t).$$

The major idea for finding the terms $S^{(i,k)}$ and the first terms of the transformed Hamiltonian $\tilde{H}^{(0,1)}+\tilde{H}^{(1,0)}+\tilde{H}^{(2,0)}+\tilde{H}^{(0,2)}+\tilde{H}^{(1,1)}$ is the absence of rapidly time-varying terms in $\tilde{H}^{(i,k)}(t)$ in the interaction representation. This requirement of algebraic perturbation theory distinguishes the approach [2-4] from other approaches to constructing the effective Hamiltonian of the theory of open quantum systems [9, 10].

As a result of conventional [2-4,7,12,14] calculations, there emerge terms in the second order of the effective Hamiltonian that are diagonal in atomic variables and do not depend on photon variables

(Lamb shifts). They can be conveniently included into the energies of atomic quantum states. Besides, there occurs an operator of the dipole-dipole interaction of atoms $V^{D-D}(t)$:

$$V^{D-D}(t) = -\int \frac{(\Gamma_1^2(\omega)+\Gamma_2^2(\omega))d\omega}{\hbar(\omega+\Omega_{21})} |d_{21}|^2 (R_- R_+ + R_+ R_- - N_a),$$

The operators $R_\pm$ and R_3 take the form:

$$R_3 = \tfrac{1}{2}\sum_s (|E_2>^{(s)}<E_2|^{(s)} - |E_1>^{(s)}<E_1|^{(s)}),\ R_- = \sum_s |E_1>^{(s)}<E_2|^{(s)},\ R_+ = \sum_s |E_2>^{(s)}<E_1|^{(s)},$$

and obey the commutation relation of the $su(2)$ algebra

$$[R_3, R_\pm] = \pm R_\pm,\ [R_+, R_-] = 2R_3.$$

Projection operators associated with the non-resonant energy levels can be omitted in the rest second-order terms $\tilde{H}^{(0,1)} + \tilde{H}^{(1,0)} + \tilde{H}^{(2,0)} + \tilde{H}^{(0,2)} + \tilde{H}^{(1,1)}$. Therefore, on the basis of the terms $\tilde{H}^{(0,1)} + \tilde{H}^{(1,0)} + \tilde{H}^{(2,0)} + \tilde{H}^{(0,2)} + \tilde{H}^{(1,1)}$ it is not difficult to derive the effective Hamiltonian in the Dirac picture

$$H^{Eff}(t) = \sum_{j=1,2} (V_j^R(t) + H_j^{St}(t)) + V^{D-D}(t) + V_{12}^{Int}(t), \tag{4}$$

where $V_j^R(t)$ is the resonant interaction operator of electromagnetic fields with an atomic ensemble

$$V_j^R(t) = -\int \Gamma_j(\omega) d\omega b_j(\omega) d_{21} e^{i\Omega_{21}t} e^{-i\omega t} R_+ + H.c.,$$

$H_j^{St}(t)$ is the operator of the Stark interaction of an atomic ensemble with the j-th vacuum field where the approximation of a localized atomic ensemble is essential:

$$H_j^{St}(t) = \int d\omega d\omega' \Gamma_j(\omega)\Gamma_j(\omega') b_j^+(\omega) b_j(\omega') e^{i(\omega-\omega')t} \sum_{i,k=1,2} \tfrac{1}{2}(\Pi_k(\omega) + \Pi_k(\omega')) |E_k>^{(i)}<E_k|^{(i)},$$

at last, operator $V_{12}^{Int}(t)$ describes the mutual interaction of broadband fields on the atoms of the ensemble:

$$V_{12}^{Int}(t) = \int d\omega d\omega' \Gamma_1(\omega)\Gamma_1(\omega') b_1^+(\omega) b_2(\omega') e^{i(\omega-\omega')t} \sum_{i,k=1,2} \tfrac{1}{2}(\Pi_k(\omega) + \Pi_k(\omega')) |E_k>^{(i)}<E_k|^{(i)} + H.c. \tag{6}$$

The letters $H.c.$ denote the term that is Hermitian conjugate to the previous one. Atomic operator $\sum_i |E_k>^{(i)}<E_k|^{(i)}$ for resonance levels is expressed in terms of generators $su(2)$ of algebra as in

$$\sum_i |E_1>^{(i)}<E_1|^{(i)} = \tfrac{1}{2}\sum_i (|E_1>^{(i)}<E_1|^{(i)} + |E_2>^{(i)}<E_2|^{(i)}) - \tfrac{1}{2}\sum_i (|E_2>^{(i)}<E_2|^{(i)} - |E_2>^{(i)}<E_2|^{(i)}),$$

$$\sum_i |E_2>^{(i)}<E_2|^{(i)} = \tfrac{1}{2}\sum_i (|E_1>^{(i)}<E_1|^{(i)} + |E_2>^{(i)}<E_2|^{(i)}) + \tfrac{1}{2}\sum_i (|E_2>^{(i)}<E_2|^{(i)} - |E_2>^{(i)}<E_2|^{(i)}).$$

According to the accepted assumption, atoms populate only resonance levels, hence

$$\sum_i |E_1>^{(i)}<E_1|^{(i)} = \frac{N_a}{2} - R_3,\ \sum_i |E_2>^{(i)}<E_2|^{(i)} = \frac{N_a}{2} + R_3.$$

In the second-order algebraic perturbation theory, there occur typical parameters $\Pi_k(\omega) = \sum_j{}' \frac{|d_{kj}|^2}{\hbar}\left(\frac{1}{\Omega_{kj}+\omega} + \frac{1}{\Omega_{kj}-\omega}\right)$, $\Omega_{km} = (E_k - E_m)/\hbar$. The prime at the summation sign means that the summation eliminates the resonance terms.

If we additionally leave only the diagonal terms in terms $H_j^{St}(t)$ with respect to the field operators, i.e. instead of $H_j^{St}(t)$ we consider the following values

$$H_j^{(2)Diag}(t) = \int d\omega \,|\Gamma_j(\omega)|^2 b_j^+(\omega) b_j(\omega) \sum_{i,k=1,2} \Pi_k(\omega)\,|E_k>^{(i)}<E_k|^{(i)},$$

then these values comply with the results of [9, 10]. However, from the point of view of algebraic perturbation theory, there are no grounds to introduce additional requirements of this kind for values of the second order of smallness in interaction constants. Moreover, if one of the fields, say, $j=2$ is replaced by a single-mode field of the cavity in the operator $V_{12}^{Int}(t)$, i.e. instead of $V_2(t)$ the following value is considered

$$V_2(t) \to V_c(t) = -g_c(c_c^+ e^{i\Omega_c t} + c_c e^{-i\Omega_c t}) \sum_{i,kj} d_{kj} e^{i\Omega_{kj} t} \,|E_k>^{(i)}<E_j|^{(i)}, \tag{8}$$

then, as a result of the algebraic perturbation theory, the pumping/relaxation operator for the photon mode of a single-mode cavity will follow from $V_{12}^{Int}(t)$:

$$V_{12}^{Int}(t) \to g_c \int d\omega \Gamma_1(\omega) b_1^+(\omega) c_c e^{i(\omega-\Omega_c)t} \sum_{i,k=1,2} \tfrac{1}{2}(\Pi_k(\omega) + \Pi_k(\Omega_c))\,|E_k>^{(i)}<E_k|^{(i)} + H.c. \tag{9}$$

The operator c_c mentioned above was the annihilation operator for the photon of the frequency cavity mode Ω_c, g_c - the parameter of the quantized field of a single-mode cavity.

Thus, the operators of the effective second-order Hamiltonian (4) agree with the known effective Hamiltonians of the problems of pumping by a broadband quantized field of a single-mode cavity [11,19].

3. Markov approximation and stochastic differential equation

The Markov conditions are formulated for the transformed states as the following:

1. At the initial instant of time, the states of the atomic ensemble $|\tilde{\Psi}^A(0)>$ and vacuum electromagnetic fields $|\tilde{\Psi}_1^F(0)>$ and $|\tilde{\Psi}_2^F(0)>$ are not correlated, i.e. the state function of the open system and the environment is factorized

$$|\tilde{\Psi}^{A+F}(0)> = |\tilde{\Psi}^A(0)> \otimes |\tilde{\Psi}_1^F(0)> \otimes |\tilde{\Psi}_2^F(0)>,$$

and the electromagnetic field modes are statistically independent of each other (the delta-correlation property of the modes). Only the following mean value is other than zero

$$<\tilde{\Psi}_j^F(0)|\, b_j(\omega) b_j^+(\omega')\,|\tilde{\Psi}_j^F(0)> = \delta(\omega-\omega').$$

2. The parameters of interaction between the atomic ensemble and the medium, determined by the values $\Gamma_j(\omega)$ and $\Pi_k(\omega)$, are independent of frequencies and are permanent values

$$\Gamma_j(\omega) = const = \Gamma_j,\; \Pi_k(\omega) = const = \Pi_k.$$

3. The limits of integration over frequencies range from $-\infty$ to $+\infty$.

Under these assumptions, the effective Hamiltonian is expressed in terms of the operator values

$$b_j(t) = \int_{-\infty}^{\infty} \frac{d\omega}{\sqrt{2\pi}} e^{-i(\omega-\Omega_{21})t} b_j(\omega),\; B_j(t) = \int_0^t dt' b_j(t'),\; dB_j(t) = B_j(t+dt) - B_j(t), \tag{10}$$

$$\Lambda_{jj'}(t) = \int_0^t dt' b_j^+(t') b_{j'}(t'),\; d\Lambda_{jj'}(t) = \Lambda_{jj'}(t+dt) - \Lambda_{jj'}(t).$$

Henceforward all values are assumed to be dimensionless; a non-dimensionalized procedure is performed according to work [12].

IOP Conf. Series: Journal of Physics: Conf. Series **1068** (2018) 012003 doi:10.1088/1742-6596/1068/1/012003

An important circumstance is the need to define integrals with respect to differentials $dB_j(t)$ and $d\Lambda_{jj'}(t)$ in terms of Ito integrals [11-13]. Then the Ito differentials $dB_j(t)$ and $d\Lambda_{jj}(t)$ that describe only one *j*-th broadband field satisfy the Hudson-Parthasarathy algebra [13]:

$$d\Lambda_{jj}(t)d\Lambda_{jj}(t)=d\Lambda_{jj}(t),\ d\Lambda_{jj}(t)dB_j^+(t)=dB_j^+(t),\ dB_j(t)d\Lambda_{jj}(t)=dB_j(t),\ dB(t)dB^+(t)=dt, \quad (11)$$

$$dB_j^+(t)dB_j(t)=dB_j(t)dB_j(t)=d\Lambda_{jj}(t)dB_j(t)=d\Lambda_{jj}(t)dt=dB_j^+(t)d\Lambda(t)=dB_j^+(t)dt=dB_j(t)dt=0,$$

$$dtdt=0,\ <dB_j(t)>=<dB_j^+(t)>=<d\Lambda_{jj}(t)>=0.$$

Angular brackets mean averaging over the initial state.

However, within the given problem, there occur quantum processes of the photon transfer $\Lambda_{12}(t)$ and $\Lambda_{21}(t)$, apart from the counting process $\Lambda_{jj}(t)$ for the *j*-th field. To obtain the relations involving these processes, we need a generalization of the Hudson-Parthasarathy algebra which is taken in the following form:

$$dB_j(t)dB_{j'}^+(t)=dt\delta_{jj'},\ d\Lambda_{j'j}(t)d\Lambda_{j'j}(t)=d\Lambda_{j'j}(t)\delta_{j'j},\ d\Lambda_{j'j}(t)d\Lambda_{jj'}(t)=d\Lambda_{j'j'}(t), \quad (12)$$

$$d\Lambda_{jj'}(t)dB_{j'}^+(t)=dB_j^+(t),\ dB_j(t)d\Lambda_{jj'}(t)=dB_{j'}(t).$$

Then the Schrödinger equation for the evolution operator $U(t)$ of the transformed vector $|\tilde{\Psi}^{A+F}(t)>=U(t)|\tilde{\Psi}^{A+F}(0)>$ can be restated in terms of the introduced quantum stochastic processes and expressed as (χ_j,κ_j,η_0 are dimensionless interaction parameters):

$$dU(t)=\{\exp[-i(V^{D-D}dt+\sum_j(Y_j^+dB_j(t)+Y_jdB_j^+(t))+$$
$$+(\frac{N_a}{2}+\eta_0R_3)(\sum_j\kappa_jd\Lambda_{jj}(t)+\sqrt{\kappa_1\kappa_2}(d\Lambda_{12}(t)+d\Lambda_{21}(t)))]-1\}U(t),\ Y_j=\chi_jR_-. \quad (13)$$

Providing that

$$\chi_1=\chi_2=\chi,\ \kappa_1=\kappa_2=\kappa, \quad (14)$$

One can introduce effective stochastic processes instead of the processes $B_j(\tau)$ and $\Lambda_{jj'}(t)$:

$$B(t)=(B_1(t)+B_2(t))/\sqrt{2},\ \Lambda(t)=[\Lambda_{11}(t)+\Lambda_{22}(t)+\Lambda_{12}(t)+\Lambda_{21}(t)]/2. \quad (15)$$

For the introduced processes we have the relations of the conventional Hudson-Parthasarathy algebra (11):

$$dB(t)dB^+(t)=dt,\ d\Lambda(t)d\Lambda(t)=d\Lambda(t),\ d\Lambda(t)dB^+(t)=dB^+(t),\ dB(t)d\Lambda(t)=dB(t).$$

All the terms in the expansion of the exponential can be fully summed up and the SDE can be written in a form that coincides with the SDE of [12] for some particular case (14) in expression (13):

$$dU(t)=-iH^{D-D}(t)dtU(t)+(Y^+\frac{Y_\Lambda^e+iY_\Lambda}{(Y_\Lambda)^2}Ydt+Y^+\frac{Y_\Lambda^e}{Y_\Lambda}dB(t)+\frac{Y_\Lambda^e}{Y_\Lambda}YdB^+(t)+Y_\Lambda^ed\Lambda(t))U(t), \quad (16)$$

$$Y=\sqrt{2}\chi R_-,\ Y_\Lambda=2\kappa(\frac{N_a}{2}+\eta_0R_3),\ Y_\Lambda^e=e^{-iY_\Lambda}-1.$$

Similar to works [4,12,14,19], equation (16) describes a non-Wiener decay of an excited atomic ensemble of identical atoms with a typical stabilization effect of excited states with respect to collective decay.

4. Conclusion

The described application of algebraic perturbation theory [2-4] in relation of the interaction of broadband fields with a quantum multilevel system has revealed the consistency of various quantum optical models obtained using the methods of algebraic perturbation theory. There are quite a number of examples of consistency, e.g. a particular case of interaction between single-mode and broadband quantized fields, which also describes the interaction of fields on the mirror of a single-mode cavity [19], can be added by models of quasiresonance interaction in the dispersion limit [20,21] and others [7,12,18-23]. They have proved the effectiveness of the problem statement and solution in the theory of open quantum systems in the effective Hamiltonian representation obtained by methods of algebraic perturbation theory, followed by the statement of the Markov conditions and the appropriate SDEs. In the second-order algebraic perturbation theory, quantum SDEs (referred to as non-Wiener SDEs) are those of the types of (13), or (16), which are both controlled by the quantum counting process and the Wiener stochastic processes. The increased role of the quantum reading process under second-order smallness of the terms of perturbation theory could be due to the counting property $d\Lambda(t)d\Lambda(t) = d\Lambda(t)$ and was proved in [12] for a localized ensemble of identical atoms. Generalization of the theory to extended medium is important. An example of such generalization is presented in [23].

Acknowledgments
This work is supported by the Russian Foundation for Basic Research (grant No. 16-02-00453a).

References

[1] Butylkin V S, Kaplan A E, Khronopulo Yu G and Yakubovich E I 1989 *Resonant Nonlinear Interactions of Light with Matter* (Berlin: Springer-Verlag)
[2] Basharov A M, Maimistov A I and Manykin E A 1983 Sov.Phys.JETP **57** 282
[3] Maimistov A I and Basharov A M 1999 *Nonlinear optical waves* (Dordrecht: Kluwer Academic)
[4] Basharov A M 2012 JETP **115** 371
[5] Krylov N M and Bogoliubov N N (1947) Introduction to Non-Linear Mechanics (Princeton: Princeton University Press)
[6] Bogoliubov N and Mitropolsky Y 1961 *Asymptotic Methods in the Theory of Non-linear Oscillations* (New York: Gordon and Breach)
[7] Basharov A M 1992 Sov. Phys. JETP **75** 611
[8] Lax M 1966 Phys. Rev. **145** 110
[9] Breuer H-P and Petruccione F 2002 *Theory of open quantum systems* (Oxford: OUP)
[10] Accardi L, Lu Y G and Volovich I 2002 *Quantum theory and its stochastic limit* (Berlin: Springer-Verlag)
[11] Gardiner C W and Zoller P 2004 *Quantum noise* (Berlin: Springer-Verlag)
[12] Basharov A M 2011 Phys.Rev. A **84** 013801.
[13] Hudson R L and Parthasarathy K R 1984 Comm.Math.Phys. **93** 301
[14] Basharov A M 2014 Optics and Spectroscopy **116** 495
[15] Wegner F and Wagner M 1994 Ann.Phys. **3** 77
[16] Glazek S D and Wilson K G 1993 Phys.Rev. D **48** 5863
[17] Glazek S D and Wilson K G 1994 Phys.Rev. D **49** 4214
[18] Basharov A M 2017 J Phys. CS **859** 012003
[19] Basharov A M 2012 Phys.Lett. A **376** 1881
[20] Basharov A M, Gorbachev V N and Rodichkina A A 2006 Phys.Rev. A **74** 042313 (2006)
[21] Basharov A M and Gorbachev V N 2007 Optics and Spectroscopy **102** 585
[22] Basharov A M 2015 J Phys. CS Series **613** 012007
[23] Basharov A M 2018 Zh.Eksp.Teor.Fiz. **153** 375

IOP Conf. Series: Journal of Physics: Conf. Series **1068** (2018) 012004 doi:10.1088/1742-6596/1068/1/012004

Investigation by MOKE and MFM of the domain structure transformation under mechanical deformations in permalloy microparticles

D A Bizyaev[1], A A Bukharaev[1,2], V V Churkov[1], A P Chuklanov[1], T A Khanipov[1], N I Nurgazizov[1]

[1] ZPTI – subdivision of FIC KazanSC of RAS, 420029, Sibirsky Trakt 10/7, Kazan, Russia

[2] Kazan Federal University, 420008, Kremlyovskaya 18, Kazan, Russia

E-mail: niazn@mail.ru

Abstract. In this work, the change in the domain structure of planar permalloy microparticles due to mechanical stress has been studied. For this purpose, an array of particles in a mechanically stressed state was formed on a silicon substrate. In addition, the samples with the array of particles without tension were manufactured. The magnetic structure of the samples was visualized by magnetic force microscopy and hysteresis loops were obtained using the magneto-optical Kerr effect. It was established that the easy magnetization axis collinear to the direction of mechanical stress appears in particles due to mechanical compression. The distribution of magnetization of unstrained particles is mainly determined by the shape anisotropy.

1. Introduction

The possibility of using the Villari effect (or the magnetoelastic effect, in which the magnetic properties of a solid change under the mechanical action) for the process of the magnetization reversal of micro- and nanostructures has been intensively studied [1-7]. Using this effect, it is possible to significantly change the value of the external magnetic field necessary for the magnetization reversal of an individual particle, which can be used as the information storage. In some cases, the process of the magnetization reversal of a particle can be performed only by its mechanical stretching or compression [3, 4]. A significant change in the magnetization of the particle due to the mechanical action can be used to detect mechanical stresses arising in the particle. In the case of planar particles located on a solid surface, the change in their magnetization makes it possible to detect mechanical stresses in the near-surface layer of the substrate.

It is possible to study the magnetic structure of an individual microparticle in detail using a magnetic force microscope (MFM). To observe the changes in the magnetic structure of an array of particles, which cover the surface uniformly, one can use the magneto-optical Kerr effect (MOKE). In this paper, these two methods were used to study the magnetic properties of planar permalloy (Py) microparticles as a function of the external mechanical stress created in them.

2. Sample preparation

The study was carried out on samples being arrays of planar Py particles placed on the Si substrate with linear dimensions of $12\times3\times0.4$ mm^3. The permalloy composition was 79% Ni, 16% Fe and 4% Mo. Each particle had dimensions of $7.5\times7.5\times0.03$ μm^3. The distance between particles was 5 μm. Using the

Si substrate doped with phosphorus makes it possible to avoid the accumulation of the electrostatic charge, which interferes with the MFM measurements. As described in our previous work [3], sputtering of permalloy through the metal mesh pressed to the substrate was used to produce an array of identical particles. Sputtering was carried out by electron beam evaporation of a solid Py rod in ultrahigh vacuum on an Omicron Multiprobe P setup.

Two types of samples were manufactured: samples with non-stressed particles and the samples with particles compressed along the long axis of the substrate. To create mechanical stress, the substrate was bent elastically in a sample holder prior to the permalloy deposition. The mesh was fixed in such a way that one of the sides of the square particle was parallel to the axis of compression. The measurements showed that mechanical stresses created in the particles remain constant for a long time (about several months), which makes it possible to study the sample by various methods. The process of substrate bending is often used to create stresses in thin ferromagnetic films or particles [2, 6, 7]. The size and shape of particles in the different parts of the sample was controlled by atomic force and optical microscopy. According to the measurements, the size and shape of the particles was the same throughout the entire sample. The uniformity control of the particles was important for carrying out MOKE measurements, when a signal from an array of particles is recorded simultaneously.

3. MOKE experiments

A scanning magneto-polarimetric (MOKE) setup based on an LEFT-3M-1 ellipsometer was used to record the hysteresis loops of the array if Py particles. The samples under investigation were rotated in the MOKE setup in the plane of the sample surface with a step of 5 degrees.

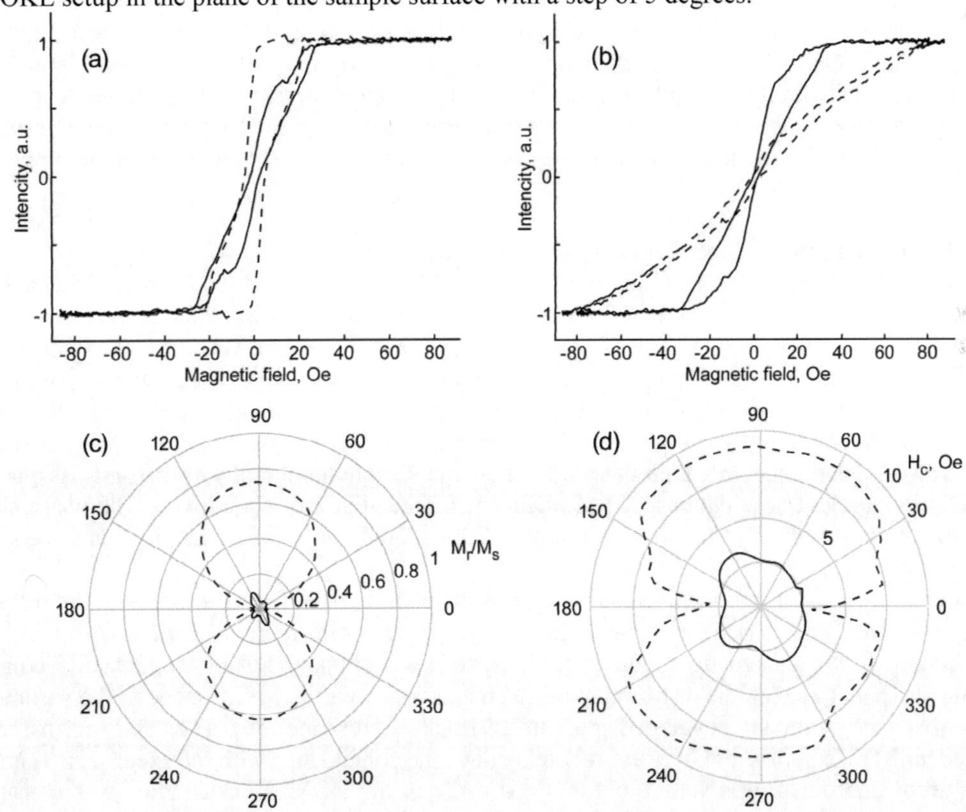

Figure 1. Hysteresis loops obtained by the MOKE technique for non-stressed (solid lines) and stressed (dashed lines) particles with the direction of the external magnetic field along the easy axis (a) and along the hard axis (b). The angular dependences of the normalized remanent magnetization (c) and the coercive field of the particles (d)

The initial orientation of the sample was parallel to one of the sides of the square particle. In addition, in the case of strained sample, the initial orientation was parallel to the axis of compression. The hysteresis loop was recorded by the MOKE technique at each rotation position from the region of the size of about 10^{-2} mm^2 (figure 1 a,b). Each point on the hysteresis loop is the result of averaging over 200 measurements. Based on the obtained hysteresis loops, the azimuthal dependences of the remanent magnetization of the sample (figure 1c) and the coercive field of the particles were calculated (figure 1d).

Figure 1c shows that there are two in-plane easy magnetization axes on the sample with non-stressed particles. The directions of the easy axis are close to the direction of diagonals of particles, and coincide well with the domain walls observed in the MFM studies. One easy axis was formed in the stressed sample with the direction parallel to the direction of compression of particles. At the same time, a hard axis is formed in the perpendicular direction. The direction of the easy axis along the particle compression is due to the negative value of the magnetostriction coefficient of permalloy under study (at the positive value of the magnetostriction coefficient the easy axis would be perpendicular to the direction of particles compression). The mechanical stress in the particles leads to the increase in the coercive field of particles in 2-3 times in all directions, except for the hard axis, where the coercive field increases by less than 2 times (figure 1d).

The external magnetic field of 80 Oe is necessary for the uniform magnetization of a stressed particle along the hard axis, while for a non-stressed particle the magnetic field of 37 Oe is sufficient. The value of the external magnetic field necessary for the uniform magnetization of the particle along the easy axis is 24 Oe for the stressed particle and 28 Oe for the non-stressed particles. The small difference in the uniform magnetization fields for non-stressed particles in easy and hard directions indicates the weak anisotropy. The MOKE data allow us to conclude that the anisotropy of the non-stressed particles is determined primarily by their geometric shape. The mechanical stress in the particles leads to the uniaxial anisotropy. In this case, the value of the external magnetic field necessary for the uniform magnetization decreases in the direction of the easy axis and significantly increases in the direction of the hard axis.

4. MFM of the samples in the external magnetic field

A Solver P47 scanning probe microscope (SPM) (NT-MDT) and a N18 Co-Cr magnetic cantilever (MikroScience) were used to perform MFM studies. In addition, the SPM was equipped with an external magnetic field source with the intensity of up to 200 Oe, which can be applied in the plane of the sample. The magnetic structure of the particle studied in the external magnetic field applied parallel and perpendicular to the axis of compression. A series of experiments on the magnetization reversal in the external field was also carried out on non-stressed particles. The MFM measurements were performed using a two-pass technique, in which the probe scans the sample topography at the first pass, and the phase shift of cantilever oscillations (MFM image) is recorded at the second pass. The phase shift is proportional to the gradient of the magnetic force of the interaction between the probe and the sample in the given point.

To determine the distribution of magnetization in the particle from the MFM images, the following approach was used. First, based on the three-dimensional image of the particle obtained on the SPM, the distribution of the local magnetic moments in the particle was simulated using the OOMMF computer program [8]. Then, based on the distribution of magnetization, a virtual MFM image of the particle was simulated by our previously developed program "Virtual microscope" [9]. The resulting image was compared with the experimental one. In order to achieve the coincidence with the experimental images, the anisotropy coefficient was varied and the next cycle of the OOMMF calculation was performed. When the simulated and experimental images coincided, the conclusion was drawn that the calculated distribution of the magnetization in the Py particle coincides with the real one.

The experimental MFM images of Py particles in the external magnetic field applied along one of the sides showed the following. In the external magnetic field with the absolute value greater than 40 Oe the non-stressed particles remain magnetized uniformly and the characteristic black-and-white contrast is observed on opposite sides of the particle (figure 2f). The particle becomes quasi-single-domain at the magnetic field in the range from 35 to 40 Oe modulo and the black-and-white contrast regions on the opposite sides of the particle increase (figure 2e). The particles are in the four-domain state at the magnetic field from 0 to 30 Oe modulo. One of the magnetic domains with the direction of magnetization parallel to the external magnetic field is much larger than the domain with the opposite direction (figure 2c).

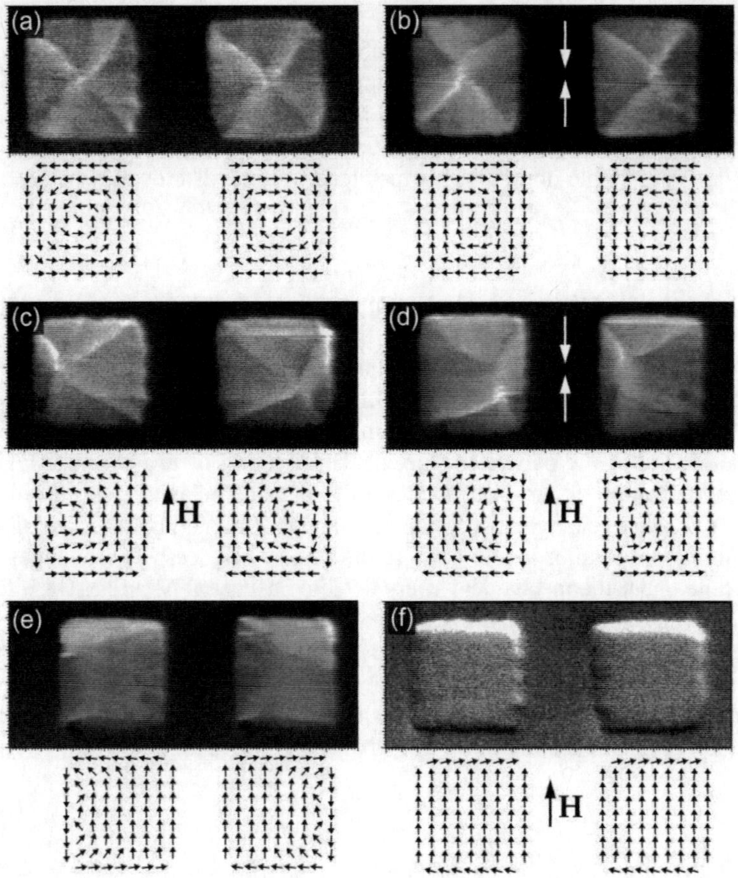

Figure 2. MFM images of two Py particles and the corresponding magnetization structure in the case of: non-stressed particles without the external magnetic field (a), stressed particles without the field (b), non-stressed particles in the field of -15 Oe (c), stressed particles in the field of -15 Oe (d), non-stressed particles (quasi-single-domain) in the field -35 Oe (e), uniformly magnetized particles in the field of -80 Oe (f). The white arrows show the direction of compression. The scanning range is 22×9 μm

For stressed particles with the direction of the external magnetic field along one of the particle sides and parallel to the axis of compression of particles (or the easy axis) the similar behavior of the magnetization was observed. The particle remains magnetized uniformly in the magnetic fields higher than 30 Oe modulo, becomes quasi-single-domain (similar to figure 2e) in the fields of 20-25 Oe, and four-domain in 0-20 Oe (figure 2d)

If the magnetic field was applied along one of particle side and perpendicular to the axis of compression (along hard axis), the behavior of a particle changes. The particle remains magnetized uniformly in the magnetic field higher than 60 Oe modulo. The quasi-single-domain state of the particle is observed at the interval of absolute values of the magnetic field of 45-55 Oe. In the fields of 0-40 Oe modulo, both the four-domain and seven-domain structures are observed. According to the MFM measurements, the higher external magnetic field is required for the uniform magnetization of the stressed particle in the direction of the hard axis, which coincides with the results observed with the MOKE technique.

The behavior of the magnetic structure of the studied Py particles in the external magnetic field is typical for multidomain particles [10]. The area of the domain with the direction of magnetization, which coincides with the direction of the magnetic field, increases with increasing magnetic field. It is due to the reduction in the size of the domain with the magnetization of which is in the opposite direction (figure 2c,d). This stage of remagnetization of particles is visualized well in the MFM measurements. After absorbing the domain with magnetization opposite to the direction of the magnetic field, the particle becomes a quasi-single-domain and its MFM image changes slightly with the increase in the external magnetic field. Therefore, it is not always possible to clearly establish the value of the field, at which the particle becomes single-domain. This may lead to some discrepancy in the MFM and MOKE results.

5. Summary

The measurements showed that the mechanical compression of the planar Py microparticles leads to the formation of the easy axis of magnetization, the direction of which coincides with the direction of compression. According to the MOKE and MFM data, the value of the external magnetic field necessary for the uniform magnetization of particles in the direction of the easy axis decreases for compressed particles. At the same time, the significant increase in the external magnetic field intensity necessary for the uniform magnetization of the particle occurs in the direction of the hard axis. This is due to the redistribution of magnetization of particles because of the magnetoelastic effect during compression of the particle. Due to the negative magnetostriction coefficient of permalloy, the size of the domains with direction of magnetization parallel to the compression axis increases. The changes in the magnetic properties of a particle under the mechanical stress detected using MFM and MOKE techniques can be used for the detection these mechanical stresses.

Acknowledgments

This work was partly supported by "Russian Foundation for Fundamental Research": the MOKE-experiments and micromagnetic simulations (RFFI No. 17-08-00915).

References

[1] Morozov A I 2014 *Phys. Solid State.* **56** 865
[2] Belyaev B A and Izotov A V 2007 *Phys. of the Solid State* **49** 1731
[3] Bizyaev D A, Bukhraev A A, Kandrashkin Yu E, Mingalieva L V, Nurgazizov N I and Khanipov T F 2016 *Technical Physics Lett.* **42(10)** 1034–37
[4] Peng R C, Wang J J, Hu J M, Chen L Q and Nan C W 2015 *Appl. Phys. Lett.* **106** 142901
[5] Finizio S, Foerster M., Buzzi M, Kruger B, Jourdan V, Vaz C A F, Hockel J, Miyawaki T, Tkach A, Valencia S *et al* 2014 *Phys. Rev. Applied.* **1** 021001
[6] Kumar D, Singh S, Vishawakarma P, Dev A S, Reddy V R and Gupta A 2016 *J. Magn. Magn. Mater.* **418** 99-106
[7] Dai G, Zhan Q, Yang H and Li R W 2013 *J. Appl. Phys.* **114** 173913
[8] Donahue M J and Porter D G. URL:http://math.nist.gov/oommf
[9] Ovchinnikov D V and Bukharaev A A 2001 *Tech. Phys.* **46** 1014
[10] Vonsovskii S V 1974 *Magnetism* (New York: J. Wiley) p 1256

Electromagnetic mass of an electron in one-dimensional photonic crystal

R Kh Gainutdinov[1, 2], **A I Garifullin**[1], **K A Ziyatdinova**[1], **M A Khamadeev**[1, 2] and **M Kh Salakhov**[1, 2]

[1] Kazan Federal University, Institute of Physics, 420008, Kremlevskaya St, 18, Kazan, Russian Federation
[2] Tatarstan Academy of Sciences, 420111, Bauman St, 20, Kazan, Russian Federation

E-mail: Renat.Gainutdinov@kpfu.ru

Abstract. Electromagnetic mass of an electron in one-dimensional photonic crystal is investigated. The exact expression describing the dependence of this mass on the angular coordinate of the electron's momentum is derived.

1. Introduction

Photonic crystals (PCs) are optical media with periodic modulation of refractive index along one, two or three directions in space [1–3]. This leads to the appearance of a forbidden frequency interval called the photonic band gap [4]. These unusual optical properties determine the wide applications of PC, such as simultaneous generation of second to fifth harmonic conical beams in a two dimensional nonlinear photonic crystal [5], creation of the physical and biochemical sensors [6], SERS experiments [7], solar cell light trapping [8] *etc*. At the same time, in Ref. [9] it has been shown that there is a novel effect in PC medium which consists in the fact that the electromagnetic mass (EMM) of an electron should change its value:

$$\delta m_{em}^{pc} = \delta m_{pc} + \delta m_{em}^{vac}, \tag{1}$$

where δm_{em}^{vac} and δm_{em}^{pc} are EMM in vacuum and total EMM in PC. They are both undefined in standard QED. But PC EMM δm_{pc} is finite and could be calculated. In Ref. [9] it has been shown that δm_{pc} depends on the structure of the electromagnetic field of the PC medium and in the case of 3D PC its numerical calculation becomes very difficult. For this reason the aim of this paper is to derive the exact formula for PC EMM in 1D PC because this medium is very simple from theoretical and practical points of view.

2. The electromagnetic field in one-dimensional photonic crystal and the electromagnetic mass of an electron

Let us consider the electromagnetic field in one-dimensional photonic crystal in the Bloch form:

$$\mathbf{E}_{\mathbf{k}n}(\mathbf{r}) = \sum_G \mathbf{E}_{\mathbf{k}n}(G,\mathbf{k}) e^{i(\mathbf{k}+G\mathbf{e}_z)\cdot\mathbf{r}}. \tag{2}$$

For this field the PC EMM of an electron takes the form [9] (here and below, we use the natural system of units in which $\hbar = c = 1$):

$$\delta m_{pc} = \frac{\alpha}{\pi^2}\left[\sum_n^{\omega_{\mathbf{k}n}<\Lambda} \int_{FBZ}^{\omega_{\mathbf{k}n}<\Lambda} \frac{d^3\mathbf{k}}{\omega_{\mathbf{k}n}^2} \sum_G \frac{\left|\mathbf{p}\cdot\mathbf{E}_{\mathbf{k}n}(G,\mathbf{k})\right|^2}{\mathbf{p}^2} - \int^{\omega_{\mathbf{k}n}<\Lambda} \frac{d^3\mathbf{k}}{2\mathbf{k}^2} \sum_{\lambda=1}^{2} \frac{\left|\mathbf{p}\cdot\boldsymbol{\varepsilon}_\lambda(\mathbf{k})\right|^2}{\mathbf{p}^2}\right], \tag{3}$$

where $G = N_1 \cdot g$ is the absolute value of the reciprocal lattice vector [10, 11], α is the fine structure constant, $\omega_{\mathbf{k}n}(\mathbf{k})$ are dispersion relations determining the dependence of the photon's energy on its momentum. Λ is a cutoff depending on dispersion properties of the medium. This parameter is used, because virtual high-energy photons emitted by an electron in the PC medium propagate as if they were in vacuum. In one-dimensional photonic crystal the electromagnetic field can be presented in the form of two mutually perpendicular vectors:

$$\mathbf{E}_{\mathbf{k}n}(G,\mathbf{k}) = E_{\mathbf{k}n}^{\perp}(G,\mathbf{k})\boldsymbol{\varepsilon}_G^{\perp} + E_{\mathbf{k}n}^{\parallel}(G,\mathbf{k})\boldsymbol{\varepsilon}_G^{\parallel}, \tag{4}$$

$\boldsymbol{\varepsilon}_G^{\perp}$ and $\boldsymbol{\varepsilon}_G^{\parallel}$ are unit vectors of the field polarization (see Figure 1).

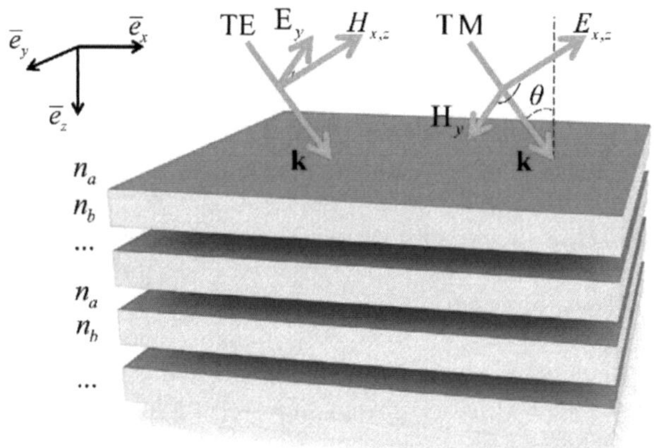

Figure 1. TE and TM polarizations of electromagnetic field in one-dimensional photonic crystal.

The dispersion relations and scalar products in (3) can be calculated by plane wave expansion method and transfer matrix method [12]. In the presence of the cutoff the electromagnetic mass of the electron in vacuum takes the form $\delta m_{em}^{vac} = \frac{4\alpha}{3\pi}\Lambda$ [13]. Using the presentation of the field in the form (4) the total EMM of the electron in PC can be written as the sum of three components:

$$\delta m_{em}^{pc} = \delta m_1 + \delta m_2 + \delta m_3, \tag{5}$$

with

$$\delta m_1 = \frac{\alpha}{\pi^2} \sum_n^{\omega_{\mathbf{k}n}<\Lambda} \int_{FBZ}^{\omega_{\mathbf{k}n}<\Lambda} \frac{d^3\mathbf{k}}{\omega_{\mathbf{k}n}^2} \sum_G \left(f_{\mathbf{p}\cdot\mathbf{E}_n^{\perp}} E_n^{\perp}(G,\mathbf{k})\right)^2, \tag{6}$$

$$\delta m_2 = \frac{\alpha}{\pi^2} \sum_n^{\omega_{\mathbf{k}n}<\Lambda} \int_{FBZ}^{\omega_{\mathbf{k}n}<\Lambda} \frac{d^3\mathbf{k}}{\omega_{\mathbf{k}n}^2} \sum_G 2\left(f_{\mathbf{p}\cdot\mathbf{E}_n^{\perp}} E_n^{\perp}(G,\mathbf{k})\right)\left(f_{\mathbf{p}\cdot\mathbf{E}_n^{\parallel}} E_n^{\parallel}(G,\mathbf{k})\right), \tag{7}$$

$$\delta m_3 = \frac{\alpha}{\pi^2}\sum_n^{\omega_{\mathbf{k}n}<\Lambda}\int_{FBZ}^{\omega_{\mathbf{k}n}<\Lambda}\frac{d^3\mathbf{k}}{\omega_{\mathbf{k}n}^2}\sum_G\left(f_{\mathbf{p}\cdot\mathbf{E}_n^{\parallel}}E_n^{\parallel}(G,\mathbf{k})\right)^2. \tag{8}$$

where $f_{\mathbf{p}\cdot\mathbf{E}}(\Omega)=\sin\Theta_E\sin\Theta\cos(\Phi-\Phi_E)+\cos\Theta_E\cos\Theta$ is the scalar product between the unit vector of electron's momentum with angular coordinates $\Omega=(\Theta,\Phi)$ and the field unit vectors $\boldsymbol{\varepsilon}_{G\parallel}$ and $\boldsymbol{\varepsilon}_{G\perp}$ with angular coordinates (Θ_E,Φ_E). After integration of expressions (6-8) in cylindrical coordinate system over the azimuthal angle φ_k, the components can be presented in the form:

$$\delta m_1 = \frac{\alpha}{\pi}\sum_n^{\omega_{\mathbf{k}n}<\Lambda}\int k_\rho dk_\rho\int_{FBZ}\frac{dk_z}{\omega_{\mathbf{k}n}^2}\sum_G E_n^{\perp 2}(G,k_\rho,k_z)\left(\frac{k_z^2}{k_\rho^2+k_z^2}\sin^2\Theta+2\frac{k_\rho^2}{k_\rho^2+k_z^2}\cos^2\Theta\right), \tag{9}$$

$$\delta m_2 = 0, \tag{10}$$

$$\delta m_3 = \frac{\alpha}{\pi}\sum_n^{\omega_{\mathbf{k}n}<\Lambda}\int k_\rho dk_\rho\int_{FBZ}dk_z\frac{1}{\omega_{\mathbf{k}n}^2}\sum_G E_n^{\parallel 2}(G,k_\rho,k_z)\sin^2\Theta, \tag{11}$$

where k_ρ, k_z are radial component and z-component of the wave vector $\mathbf{k}$ in cylindrical coordinate system. Then the total electromagnetic mass of the electron in one-dimensional photonic crystal is equal:

$$\delta m_{em}^{pc} = \frac{\alpha}{\pi}\sum_n^{\omega_{\mathbf{k}n}<\Lambda}\int k_\rho dk_\rho\int_{FBZ}\frac{dk_z}{\omega_{\mathbf{k}n}^2}\sum_G\left\{\frac{E_n^{\perp 2}(G,k_\rho,k_z)}{k_\rho^2+k_z^2}\left(k_z^2\sin^2\Theta+2k_\rho^2\cos^2\Theta\right)+E_n^{\parallel 2}(G,k_\rho,k_z)\sin^2\Theta\right\}. \tag{12}$$

Thus, operator (3) can be presented in the form:

$$\delta m_{pc} = A + B\cos^2\Theta, \tag{13}$$

where

$$A = \frac{\alpha}{\pi}\sum_n^{\omega_{\mathbf{k}n}<\Lambda}\int k_\rho dk_\rho\int_{FBZ}\frac{dk_z}{\omega_{\mathbf{k}n}^2}\sum_G\left(E_n^{\perp 2}(G,k_\rho,k_z)\frac{k_z^2}{k_\rho^2+k_z^2}+E_n^{\parallel 2}(G,k_\rho,k_z)\right)-\frac{4\alpha}{3\pi}\Lambda, \tag{14}$$

$$B = \frac{\alpha}{\pi}\sum_n^{\omega_{\mathbf{k}n}<\Lambda}\int k_\rho dk_\rho\int_{FBZ}\frac{dk_z}{\omega_{\mathbf{k}n}^2}\sum_G\left(\frac{E_n^{\perp 2}(G,k_\rho,k_z)}{k_\rho^2+k_z^2}\left(2k_\rho^2-k_z^2\right)-E_n^{\parallel 2}(G,k_\rho,k_z)\right). \tag{15}$$

The coefficients A and B can be calculated numerically. Formula (13) describes PC EMM of an electron propagating in the voids of 1D PC. Thus, its physical mass, will be determined by:

$$m_{pc} = m_e + \delta m_{pc}. \tag{16}$$

In the case of atomic electron PC EMM (13) becomes an operator whose diagonal matrix elements describe the shifts of energy levels.

3. Conclusion

We have shown that the electromagnetic mass of a free electron placed in one-dimensional photonic crystal depends on the angular coordinate of its momentum. The exact expression describing this dependence is derived. This correction leads to modifying of the physical mass of an electron and, hence, could be observed. It could manifest itself in shifts of energy levels of atoms placed in one-dimensional PC and in alteration of chemical bounds. The presented results could be used for precise adjustment of the energy levels of atoms and, as a consequence, for creation of new line spectrum sources.

IOP Conf. Series: Journal of Physics: Conf. Series **1068** (2018) 012005 doi:10.1088/1742-6596/1068/1/012005

References

[1] Yablonovitch E 1987 *Phys. Rev. Lett.* **58** 2059
[2] John S 1987 *Phys. Rev. Lett.* **58** 2486
[3] Joannopoulos J D, Johnson S G, Winn J N and Meade R D 2008 *Photonic Crystals: Molding the Flow of Light* (Princeton: Princeton University Press) p 286
[4] Lopez C 2003 *Adv. Mater.* **46** 1679–1704
[5] Mateos L, Molina P, Galisteo J, López C, Bausá L E and Ramírez M O 2012 *Opt. Express* **20** 29940–8
[6] Pinto A M R and Lopez-Amo M 2012 *J. Sens.* **2012** 598178
[7] Tuyen L D, Liu A C, Huang C-C, Tsai P-C, Lin J H, Wu C-W, Chau L-K, Yang T S, Minh L Q, Kan H-C and Hsu C C 2012 *Opt. Express* **20** 29266–75
[8] Callahan D M, Munday J N and Atwater H A 2012 *Nano Lett.* **12** 214–8
[9] Gainutdinov R Kh, Khamadeev M A and Salakhov M Kh 2012 *Phys. Rev.* A **85** 053836
[10] Kittel C 1996 *Introduction to Solid State Physics* (New York: Wiley Press) p 680
[11] Sakoda K 2001 *Optical Properties of Photonic Crystals* (Berlin: Springer) p 232
[12] Skorobogatiy M and Yang J 2009 *Fundamentals of Photonic Crystal Guiding* (New York: Cambridge University Press) p 267
[13] Schweber S S 2011 *An Introduction to Relativistic Quantum Field Theory* (New York: Courier) p 944

Control of emission spectrum of a quantum dot coupled to phonon and electron reservoirs

R Kh Gainutdinov[1,2], A Shirdelhavar[1], A A Mutygullina[1], M A Khamadeev[1,2]

[1]Kazan Federal University, Institute of Physics, 420008, Kremlevskaya St, 18, Kazan, Russian Federation

[2]Tatarstan Academy of Sciences, 420111, Bauman St, 20, Kazan, Russian Federation

E-mail: Renat.Gainutdinov@kpfu.ru

Abstract. The influence of phonon and electron reservoirs on the spectral emission from a quantum-dot cavity system is investigated. The line shapes strongly depend on the energy dependence of the exciton self-energy function, that in turn depends on the tunneling rate. This allows one to control the light emission from a single quantum dot.

1. Introduction

Quantum information technology requires the development of light sources, in which the photon number can be carefully controlled. Single photons can be generated through laser excitation [1–5]. However, for practical applications the electrical excitations are more preferable, and electro-luminescence from a single quantum dot (QD) within the intrinsic region of a p-i-n junction has been shown to act as an electrically driven single-photon source [6]. Quantum dots are embedded in a surrounding solid, and the carriers confined to the dot interact with their environment most notably with phonons. To control the spectral emission from exciton quantum dots the system can be coupled also to the electron reservoir. The additional interaction with the electron reservoirs can have a significant effect on the self-energy function of the semiconductor QD and its emission. In this paper, we show that the total self-energy function depends on the tunneling rate which by itself depends on the temperature, highness and thickness of the barrier, energy levels of the QD and the chemical potential of the lead. At the fixed thickness of the barrier between the QD and the lead, the tunneling rate Γ can be changed by tuning the voltage on the gate. This means that the spectral emission from the QD can be controlled.

2. Method

In our investigation of the the problem, we use the generalized dynamical equation (GDE), which has been derived in [7] as the most general equation of motion consistent with the current concepts of quantum physics. This equation turned out an important tool for solving many problems in nuclear physics [8–10], quantum optics [11] and atomic physics [12]. In the energy representation [13–17] the generalized dynamical equation takes the form

$$\frac{dT(z)}{dz} = -T(z)G_0^2(z)T(z), \tag{1}$$

IOP Conf. Series: Journal of Physics: Conf. Series **1068** (2018) 012006 doi:10.1088/1742-6596/1068/1/012006

where the operator $T(z)$ is defined as

$$T(z) = i \int d\tau \ \exp(iH_0 t_2)\tilde{S}(t_2, t_1) \exp(-iH_0 t_1), \tag{2}$$

and $G_0(z)$ is the "free" Green operator. Here $\tilde{S}(t_2, t_1)$ describes a contribution to the evolution operator from the processes in which the interaction begins at time t_1 and ends at time t_2, and H_0 is the free Hamiltonian. Equation (1) with the boundary condition

$$T(z) \xrightarrow[|z| \to 0]{} B(z), \tag{3}$$

where

$$B(z) = i \int d\tau \exp(iz\tau) \exp(iH_0 t) H_{int}(t_2, t_1) \exp(-iH_0 t_1), \tag{4}$$

and $H_{int}(t_2, t_1)$ is the interaction operator governing the dynamic of a quantum system. The Green operator $G(z)$ is determined as

$$G(z) = G_0(z) + G_0(z)T(z)G_0(z). \tag{5}$$

The contribution to the Green operator $G(z)$, which comes from the processes, has the same structure as the free Green operator $G_0(z)$. For this reason it is natural to replace $G_0(z)$ by the operator $\tilde{G}_0(z)$ describing the evolution of particles, that propagate freely or interacts only with vacuum and hence has the structure

$$\langle m_2|\tilde{G}_0(z)|m_1\rangle = \langle m_2|m_1\rangle \Big(z - E_{m_1} - C_{m_1}(z)\Big)^{-1}, \tag{6}$$

where $|m\rangle$ are the eigenvectors of the free Hamiltonian ($H_0|m\rangle = E_m|m\rangle$). Correspondingly, the operator $T(z)$ should be replaced by the operator $M(z)$ describing the interactions between particles. These operators are related as follows:

$$G_0(z) + G_0(z)T(z)G_0(z) = G(z) = \tilde{G}_0(z) + \tilde{G}_0(z)M(z)\tilde{G}_0(z). \tag{7}$$

From this relationship one can derive the equations for the self-energy function $C_n(z)$ and the interaction operator $M(z)$. For example, the equation for $C_n(z)$ reads

$$\frac{dC_n(z)}{dz} = -\sum_m \frac{\langle n|M(z)|m\rangle\langle m|M(z)|n\rangle}{\Big(z - E_m - C_m(z)\Big)^2}. \tag{8}$$

The boundary conditions for $M(z)$ and $C_n(z)$ are as follows

$$M(z) \xrightarrow[|z| \to \infty]{} B_r(z), \tag{9}$$

and

$$C_n(z) \xrightarrow[|z| \to \infty]{} B_\delta(z), \tag{10}$$

where $B_r(z)$ is the part of the interaction operator describing the proper interaction between particles, and $B_\delta(z)$ describes their self-interaction: $B(z) = B_r(z) + B_\delta(z)$.

3. Emission from exciton quantum dots coupled to phonon and electron reservoirs

Let us consider the interaction of a QD with its environment, which consists of the reservoir of longitudinal acoustic phonons and two leads being electron reservoirs. Let $|\mu\rangle$ be the eigenstate of the exciton-reservoirs system with the energy $E = E + E_\mu$. The eigenstates $|\mu\rangle$ with the energy E_μ describe the environment as a system with almost infinite number degrees of freedom, and hence the change in energy of the states $|\mu\rangle$ caused by the interaction of the QD with the reservoirs is extremely small compared to the energy of these states. For this reason, it is convenient to set E_μ to be the zero energy point. The zero point for z is changed correspondingly $z' = z - E_\mu \longrightarrow z$. By interacting with the reservoirs, QD creates and destroys particles in the reservoir. The equation for the self-energy function of the QD coupled to the phonon and electron reservoirs takes the form.

$$\frac{dC_{X\mu}}{dz} = -\sum_j \frac{\langle X, \mu\,|M\,(z)|\,j, \nu\rangle\,\langle j, \nu\,|M\,(z)|\,X, \mu\rangle}{(z - E_j - E_\nu - C_{X\mu}\,(z - E_\nu))^2}. \tag{11}$$

We will restrict ourselves to the case where only one phonon or electron is created or destroyed in the process of the interaction. In this way, we derive the equation

$$\begin{aligned}\frac{dC_{X\mu}(z)}{dz} = &-\sum\left\{\frac{\langle X;\mu\,|M(z)|\,X;\mu,+\mathbf{1}_q\rangle\,\langle X;\mu,+\mathbf{1}_q\,|M(z)|\,X;\mu\rangle}{\Big(z - E_X - \omega(q) - C_{X\mu}(z-\omega(q))\Big)^2}+\right.\\ &\left.+\frac{\langle X;\mu\,|M(z)|\,X;\mu,-\mathbf{1}_q\rangle\,\langle X;\mu,-\mathbf{1}_q\,|M(z)|\,X;\mu\rangle}{\Big(z - E_X - \omega(q) - C_{X\mu}(z-\omega(q))\Big)^2}\right\}\\ &-\sum_{\alpha,k,\sigma}\left\{\frac{\langle X;\mu\,|M(z)|\,X;\mu,+\mathbf{1}_{\alpha,k,\sigma}\rangle\,\langle X;\mu,+\mathbf{1}_{\alpha,k,\sigma}\,|M\,(z)|\,X;\mu\rangle}{\Big(z - E_X - E_k - C_{X\mu}(z-E_k)\Big)^2}+\right.\\ &\left.+\frac{\langle X;\mu\,|M(z)|\,X;\mu,-\mathbf{1}_{\alpha,k,\sigma}\rangle\,\langle X;\mu,-\mathbf{1}_{\alpha,k,\sigma}\,|M(z)|\,X;\mu\rangle}{\Big(z - E_X - E_k - C_{X\mu}(z-E_k)\Big)^2}\right\}.\end{aligned} \tag{12}$$

Where the symbol $+\mathbf{1}_q\,(-\mathbf{1}_q)$ indicates the creation (destruction) of a phonon with the momentum q in the phonon reservoir, while the symbol $+\mathbf{1}_{k\sigma\alpha}\,(-\mathbf{1}_{k\sigma\alpha})$ indicates the creation (destruction) of the electron with momentum k and the spin $\sigma \uparrow$ $(\sigma = -1)$ in lead $\alpha = L, R$, for left and right leads respectively. Here we have accounted for the fact that tunneling rate of the electron is much larger than that of the hole, and, as a consequence, the main contribution comes from the electron fluctuations between the QD and the electron reservoir. The boundary condition for this equation is as follows

$$C_{X\mu(z)} \xrightarrow[|z|\to\infty]{} 0. \tag{13}$$

The boundary condition for the operator $M(z)$ is determined by Eq. 13, and the dynamics of the quantum system is governed by the behavior of the generalized interaction operator $B(z)$ in the limit $|z| \to \infty$. If this operator is independent on z in this limit, then the dynamics is Hamiltonian and $B(z)$ is just the interaction Hamiltonian. Otherwise dynamics is non-Hamiltonian and is governed by a nonlocal-in-time interaction. It is used to start with an interaction Hamiltonian, that in the case under study takes the form

$$H_I = H_{xq} + H_{tunnel}, \tag{14}$$

IOP Conf. Series: Journal of Physics: Conf. Series **1068** (2018) 012006 doi:10.1088/1742-6596/1068/1/012006

where H_{xq} describes the interaction of the QD with the reservoir of longitudinal acoustic phonons [18]

$$H_{xq} = \sum_q g_q^x \left(b_q + b_q^\dagger |x\rangle\langle x| \right), \tag{15}$$

where $|x\rangle$ being the vector of the excitonic state, and $b_q(b_q^\dagger)$ is operator of creation (annihilation) of phonons with momentum q and energy ω_q, g_q^x is electron-phonon coupling element. The processes of tunneling between the QD and the leads is described by the second term in the right-hand part of (14) that can be spliced into two different types of tunneling H_{tun}^1 and H_{tun}^2:

$$H_{tun} = H_{tun}^1 + H_{tun}^2 \,. \tag{16}$$

The Hamiltonian H_{tun} generates couplings between different states $|n\rangle$ and $|n'\rangle$ of the global system. The Hamiltonian H_{tun}^1 describes on-shell transitions responsible for the dissipation processes and can be written as

$$H_{tun}^1 = \sum_{nn'} V_{n'n} |n'\rangle\langle n| \delta_{E_n, E_n'} \,, \tag{17}$$

where $V_{n'n}$ is the tunnel matrix element, while H_{tun}^2 describes off-shell transitions, that generate quantum fluctuations and can be represented as

$$H_{tun}^2 = \sum_{nn'} V_{n'n} |n'\rangle\langle n| (1 - \delta_{E_n, E_n'}). \tag{18}$$

The rates for classical(on-shell) transitions between states $|n\rangle$ and $|n'\rangle$ are determined by the Fermi golden rule

$$\omega_{n'n} = \frac{2\pi}{\hbar} |V_{n'n}|^2 \delta_{E_n, E_{n'}}. \tag{19}$$

Assuming that the density of states ρ_α in the leads is constant, we get the tunnel coupling strength Γ_α as

$$\Gamma_\alpha = 2\pi \rho_\alpha |V_\alpha|^2. \tag{20}$$

The remarkable feature of Eq. 12 is that it can be solved beyond the perturbation theory. However, in this work, we will restrict ourselves to the second Born approximation. In this approximation

$$M(z) = H_{qx}, \tag{21}$$

and we neglect the self-energy in the denominators on the right-hand part of Eq.12. In this way we can easily derive a solution for the self-energy function, which after thermal averaging takes the form

$$C_X(z) = \sum_q \left\{ \frac{|g_x(q)|^2 (1 + n(q))}{z - E_x - \omega(q) + i0} + \frac{|g_x(q)|^2 n(q)}{z - E_x + \omega(q) + i0} \right\} + \\ + \sum_\alpha \frac{\Gamma_\alpha}{2\pi} \int d\omega \left(\frac{-f_\alpha(\omega)}{z - w} + \frac{f_\alpha(\omega)}{z - 2\epsilon - U - \omega} \right), \tag{22}$$

where $n(q) = [\exp(\beta\omega(q)) - 1]^{-1}$ and $f_\alpha = [\exp(\beta(\omega - \mu_\alpha) + 1]^{-1}$, while $\beta = 1/k_B T$ and Γ_α is tunnel coupling strength of electrons. Let us consider emission from a single QD embedded in a high-quality cavity. In our investigations, we use an expression derived in [18] for the linear quantum light spectrum, including the influence from both exciton-and cavity-emission decay channels. This expression reads

$$S_r(\mathbf{r}, z) = F_r(\mathbf{r})\Gamma_{rad} \times \left| \frac{\omega_x + z}{\omega_x^2 - z^2 - i\omega\Gamma_x - \omega C_X(z) - \frac{4g^2\omega_x\omega_c}{\omega_c^2 - z^2 - i\omega\Gamma_c}} \right|^2, \tag{23}$$

$$S_c(\mathbf{r}, z) = F_c(\mathbf{r})\Gamma_c \times \left| \frac{\frac{2g\omega_c(\omega_x+z)}{\omega_c^2 - z - iz\Gamma_c}}{\omega_x^2 - z^2 - i\omega\Gamma_x - \omega C_X(z) - \frac{4g^2\omega_x\omega_c}{\omega_c^2 - z^2 - iz\Gamma_c}} \right|^2, \tag{24}$$

where $S_z(\mathbf{r}, z)$ and $S_c(\mathbf{r}, z)$ are the radiation-mode-emitted spectrum and cavity-emitted spectrum respectively, and $F_r(\mathbf{r})$ and $F_c(\mathbf{r})$ represent the geometrical factors that depend on the collection geometry of the emitted light. The parameter Γ_c, $\Gamma_c \equiv \Gamma_{rad} + \Gamma_d$ with Γ_d being the decay rate associated with the pure dephasing processes are included phenomenologically. In Fig. 1 we show that coupling of the exciton-boson-reservoir system to the electron reservoir has a significant effect on the total spectra from both cavity emission and the QD emission.

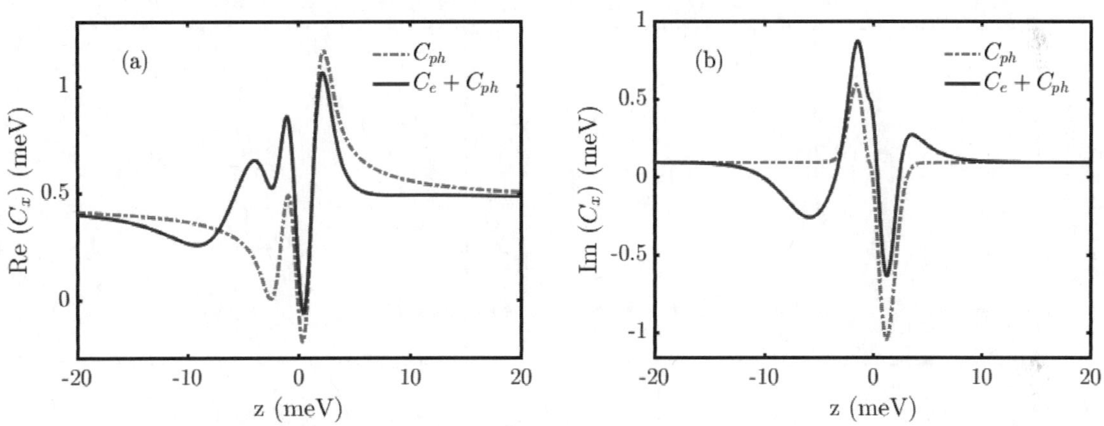

Figure 1. Exciton self-energy function ($T = 20\ K$, $U = 5\ meV$, $S_{HR} = 0.2$ and $\Gamma = 2\ meV$): (a) Real part and (b) Imaginary part. The dashed curve represents the contribution from phonon reservoir, the solid curve represents the contribution from both electron and phonon reservoirs.

IOP Conf. Series: Journal of Physics: Conf. Series **1068** (2018) 012006 doi:10.1088/1742-6596/1068/1/012006

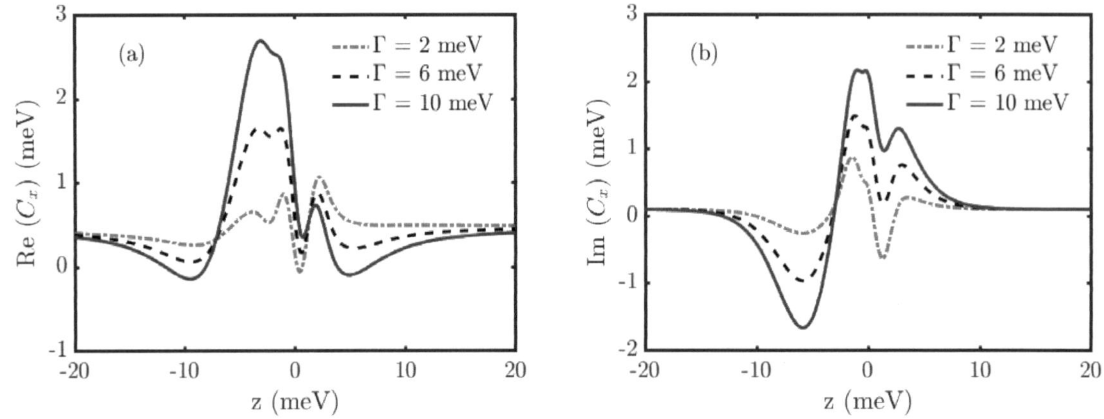

Figure 2. Exciton self-energy function ($T = 20\ K$, $U = 5\ meV$ and $S_{HR} = 0.2$): (a) Real part and (b) Imaginary part for various tunnel coupling strengths (Γ).

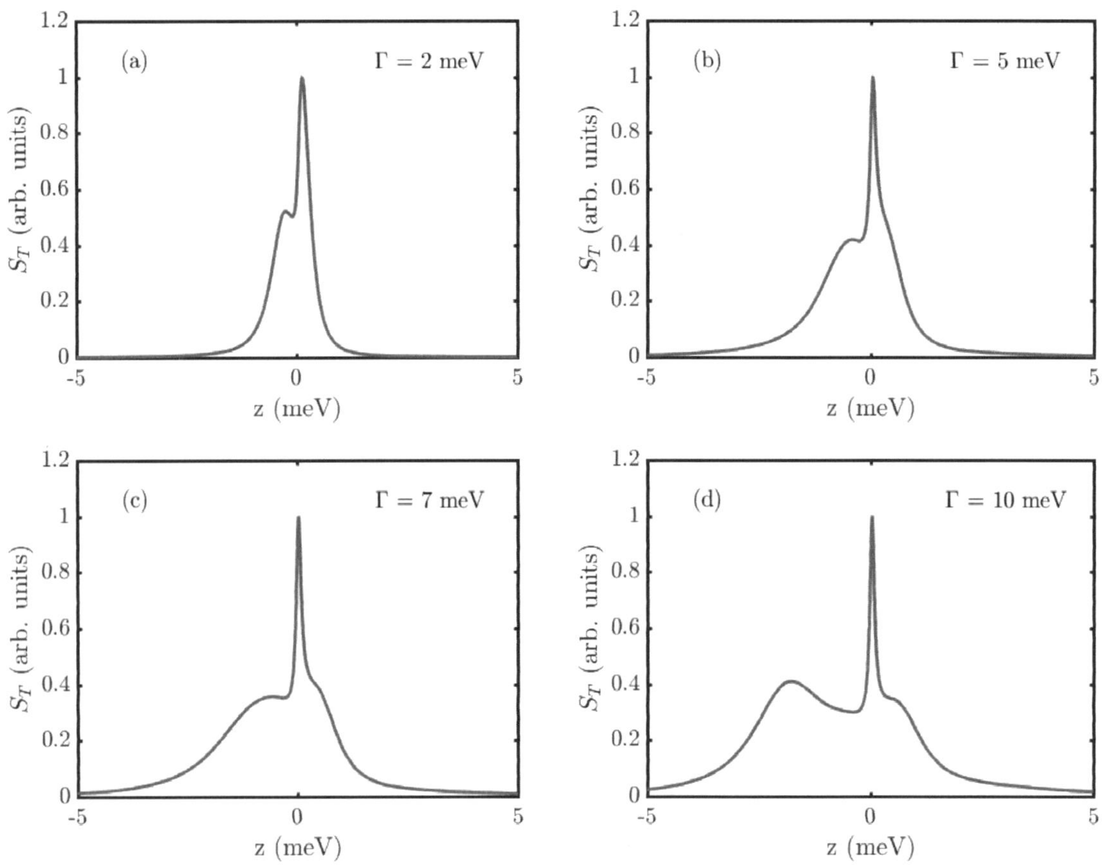

Figure 3. Radiation-spectra for various tunnel coupling strengths (a) $\Gamma = 2\ meV$, (b) $\Gamma = 5\ meV$, (c) $\Gamma = 7\ meV$, (d) $\Gamma = 10\ meV$. The parameters are $T = 20\ K$, $U = 5\ meV$ and $S_{HR} = 0.2$.

4. Conclusion

In conclusion, we have shown that in the case when in addition to the phonon reservoir the single QD is coupled to the electron reservoirs the spectral lines shapes are significantly modified. Results of calculations presented in Fig. (1-3) show that the effect significantly depends on the tunneling rate Γ. The reason for this is that the line shapes strongly depend on the energy dependence of the exciton self-energy function of $C_x(z)$, that in turn depends on the tunneling rate. These rates can be changed by tuning the voltage on the gate and in this way, the specral emission from the QD can be controlled.

References

[1] Niwa S i, Eswaramoorthy M, Nair J, Raj A, Itoh N, Shoji H, Namba T and Mizukami F 2002 *Science* **295** 105–107
[2] Brassard G, Lütkenhaus N, Mor T and Sanders B C 2000 *Physical Review Letters* **85** 1330
[3] Michler P, Imamoğlu A, Mason M, Carson P, Strouse G and Buratto S 2000 *Nature* **406** 968
[4] Michler P, Kiraz A, Becher C, Schoenfeld W, Petroff P, Zhang L, Hu E and Imamoglu A 2000 *science* **290** 2282–2285
[5] Santori C, Pelton M, Solomon G, Dale Y and Yamamoto Y 2001 *Physical Review Letters* **86** 1502
[6] Thompson R, Stevenson R, Shields A, Farrer I, Lobo C, Ritchie D, Leadbeater M and Pepper M 2001 *Physical Review B* **64** 201302
[7] Gainutdinov R K 1999 *Journal of Physics A: Mathematical and General* **32** 5657
[8] Gainutdinov R K and Mutygullina A A 2002 *Physical Review C* **66** 014006
[9] Gainutdinov R K and Moutygoullina A 1997 *Physics of Atomic Nuclei* **60** 841–847
[10] Gainutdinov R K and Mutygullina A A 2002 *Physics of Atomic Nuclei* **65** 1421–1430
[11] Gainutdinov R K and Mutygullina A 2009 *Physica Scripta* **2009** 014021
[12] Gainutdinov R K 1989 *Journal of Physics A: Mathematical and General* **22** 269
[13] Gainutdinov R K, Mutygullina A A and Scheid W 2002 *Physics Letters A* **306** 1–9
[14] Gainutdinov R K 1983 *Sov. J. Nucl. Phys.* **37** 277–282
[15] Gainutdinov R K 1987 *Sov. J. Nucl. Phys.* **46** 743–749
[16] Gainutdinov R K 1987 *Sov. J. Nucl. Phys.* **53** 885–892
[17] Gainutdinov R K 1995 *Zh. Eksp. Teor. Fis.* **108** 1600–1613
[18] Hughes S, Yao P, Milde F, Knorr A, Dalacu D, Mnaymneh K, Sazonova V, Poole P, Aers G, Lapointe J *et al.* 2011 *Physical Review B* **83** 165313

Nonperturbative effects in self-energy functions of quantum dots coupled to an acoustic phonon reservoir

R Kh Gainutdinov[1,2], D G Blum[1], A A Mutygullina[1]

[1]Kazan Federal University, Institute of Physics, 420008, Kremlevskaya St, 18, Kazan, Russian Federation

[2]Tatarstan Academy of Sciences, 420111, Bauman St, 20, Kazan, Russian Federation

E-mail: `Renat.Gainutdinov@kpfu.ru`

Abstract. The problem of description of the strong interaction of a quantum dot (QD) with an acoustic phonon reservoir is discussed. It is shown that the generalized dynamical equation of motion derived in [Gainutdinov R. Kh. 1989 *J. Phys. A: Math. Gen.* **22** 269] as the most general equation of motion consistent with the current concepts of quantum physics and provides an effective way to solve the problem of the QD in the excitonic regime with the phonon reservoir beyond the perturbation theory. A nonperturbative solution for the exciton self-energy function was found. At temperatures, more than 10 K these self-energy functions differ profoundly from that obtained in the second-Born approximation.

1. Introduction

Semiconductor quantum dots have unique optoelectronic properties [1], which make them attractive as candidates for many applications such as new types of single-photon sources [2–4], quantum repeaters [4–9] and computation building blocks of quantum computers [10–14]. Although, semiconductor quantum dots are embedded in a surrounding solid and, in general, the interaction is very strong and gives rise to the dephasing, which presents a serious challenge for practical realization of quantum computers. Also, the strong interaction of the quantum dot expresses itself in quantum fluctuations by being reversible processes in which reservoir degrees of freedom demonstrate themselves in virtual intermediate states. In each act of fluctuation, the system could go to other state and come back to the initial state. Quantum fluctuations are described by the self-energy function. The real part of this function determines the self-energy of the QD states known as the polaron shift, and it is especially important that the energy dependence of the self-energy functions have a significant effect on the emission spectra of a strongly coupled QD-cavity systems. These energy functions are exhibited their non-Markovian character of the dynamics of QDs coupled to a reservoir. In other words, in this case, nonlocality in time of the effective interaction in the system manifests itself explicitly, because an effective interaction in reduced systems is, in general, nonlocal both in space and in time.

A straightforward and effective way to describe nonlocal in time interactions is provided by a generalized dynamic equation (GDE) derived by Gainutdinov [15] as the most general equation of motion, consistent with the current concepts of quantum physics. This equation turned out an important tool for solving many problems in nuclear physics [16–18], quantum

IOP Conf. Series: Journal of Physics: Conf. Series **1068** (2018) 012007 doi:10.1088/1742-6596/1068/1/012007

optics [19] and atomic physics [20]. In this work, we show that GDE provides an effective way to solve the problem of the exciton-phonon interaction beyond the perturbation theory, and the nonperturbative solution for the self-energy function of the QD in the exciton regime differ profoundly from that obtained in the second-Born approximation.

2. Method

In the energy representation, Gainutdinov equation takes form [21–25]:

$$\frac{dT(z)}{dz} = -T(z)(G_0(z))^2T(z), \tag{1}$$

where $G_0(z) = (z - H_0)^{-1}$ with H_0 being the free Hamiltonian, and T(z) is defined as

$$T(z) = i\int_0^\infty d\tau \exp(i(z - H_0)t_2)\tilde{S}(t_2, t_1)\exp(-i(z - H_0)t_1), \tag{2}$$

with $\tilde{S}(t_2, t_1)$ being the contribution to the evolution operator from the processes, in which the interaction in the system begins at time t_1 and ends in time t_2, extends quantum dynamics to the case where dynamics of a system is governed by nonlocal in time interaction. Peculiarities of interaction in quantum systems are contained in the boundary condition for Eq.2. In particular, this fact is very important in solving various problems of many-particle physics. Operator $T(z)$ determines the Green operator

$$G(z) = G_0(z) + G_0(z)T(z)G_0(z). \tag{3}$$

The contribution to the Green operator $G(z)$, which comes from he processes associated with the self-interaction of the system, has the same structure as the free Green operator $G_0(z)$. For this reason it is natural to replace $G_0(z)$ by the operator $\tilde{G}_0(z)$,which describes the evolution of the system in the case where the interaction in the system is reduced to the self-interaction and hence has the structure

$$\tilde{G}_0(z) = (z - H_0 - C(z))^{-1}, \tag{4}$$

where the operator $C(z)$ has the same eigenvectors as $H_0(C(z)|n\rangle = C_n(z)|n\rangle, H_0|n\rangle = E_n|n\rangle)$. Correspondingly, the operator $T(z)$ is the operator describing the whole interaction in the system except the self-interaction. These operators are related as follows

$$G(z) = G_0(z) + G_0(z)T(z)G_0(z) = \tilde{G}_0(z) + \tilde{G}_0(z)M(z)\tilde{G}_0(z). \tag{5}$$

By making use of this representation, equation of motion (1) can be represented in the form of two equations, for $M(z)$ and $C(z)$. The equation for $C_n(z)$, that is of central importance for the problem under study, reads

$$\frac{d\langle n|C(z)|n\rangle}{dz} = -\sum_m \frac{\langle n|M(z)|m\rangle\langle m|M(z)|n\rangle}{(z - E_m - C_m(z))^2}. \tag{6}$$

3. The exciton self-energy in a nonperturbative regime

At temperatures higher than 10 K the interaction of a QD with its surroundings becomes strong, and, as a consequence, the perturbation theory becomes inapplicable. The surroundings practically consist of an infinite number of degrees of freedom, which act together as a whole identity referred to as reservoir. The remarkable feature of equation of motion (1) is that its form does not depend on the peculiarities of the interaction. This makes it possible to find formal solutions for various physical problems, which also do not depend on the peculiarities

IOP Conf. Series: Journal of Physics: Conf. Series **1068** (2018) 012007 doi:10.1088/1742-6596/1068/1/012007

of interaction. The equation for the self-energy function of a quantum dot in excitonic regime coupled to phonon reservoir takes the form

$$\frac{dC_{X\mu}(z)}{dz} = -\sum_{\mu'} \frac{\langle x;\mu\,|M(z)|\,x;\mu'\rangle\,\langle x;\mu'\,|M(z)|\,x;\mu\rangle}{(z-E_X-E_{\mu'}-C_{X\mu}(z))^2}, \tag{7}$$

with $|\mu\rangle$ being the eigenstates of the Hamiltonian H_R of the reservoir with eigenenergy E_μ. Since the reservoir is practically consists of an infinite number of degrees of freedom the change in the energy of the state of the QD in the excitonic regime $|X,\mu\rangle$ caused by reservoir-QD interaction is extremely small compared to the energy of these states. For this reason, it is convenient to set $E_{i\mu}$ to be the zero energy point. The zero point for z is changed correspondingly $z' = z - E_{X\mu} \to z$. The reservoir-QD interaction creates and destroys particles in the reservoir. We will restrict ourselves to the case where only one phonon is created and destroyed in the interaction process. In this way we can derive the equation

$$\frac{dC_{X\mu}(z)}{dz} = \sum_{q} \frac{\langle x;\mu\,|M(z)|\,x;\mu';+\mathbf{q}\rangle\,\langle x;\mu';+\mathbf{q}\,|M(z)|\,x;\mu\rangle}{(z-E_X-\omega(q)-C_{X\mu}(z-\omega(q)))^2} + \\ + \frac{\langle x;\mu\,|M(z)|\,x;\mu';-\mathbf{q}\rangle\,\langle x;\mu';-\mathbf{q}\,|M(z)|\,x;\mu\rangle}{(z-E_X+\omega(q)-C_{X\mu}(z+\omega(q)))^2}. \tag{8}$$

The boundary condition for this equation is

$$C_{X\mu} \to 0. \tag{9}$$

The corresponding boundary condition for the interaction operator M(z) reads

$$M(z) \to H_{qx}, \tag{10}$$

where H_{qx} is the exciton-phonon interaction Hamiltonian. In the independent boson model (IBM) H_{qx} takes the form

$$H_{qx} = \sum_q g_q^x (b_q + b_q^+), \tag{11}$$

where $|x\rangle$ and $|\mu\rangle$ are states of a QD exciton and an eigenstate of the reservoir respectively, q denotes the different phonon modes with energy ω_q the creation (b_q^+) and annihilation (b_q) operators of phonons with momentum q and frequency ω_q obey the usual commutation relations for bosons, and g_q^x is the deformation potential coupling which depends on the material parameters of the host semiconductor and the exciton wave function. In a first approximation, one can set M(z) to be equal to H_{qx}. In this way, we obtain the equation

$$\frac{dC(z)}{dz} = -\sum_{q} \frac{\langle x;\mu\,|H_{qx}|\,x;\mu';+1\mathbf{q}\rangle\,\langle x;\mu';+1\mathbf{q}\,|H_{qx}|\,x;\mu\rangle}{(z-E_X-\omega(x)-C_{X\mu}(z-\omega(q)))^2} + \\ + \frac{\langle x;\mu\,|H_{qx}|\,x;\mu';-1\mathbf{q}\rangle\,\langle x;\mu';-1\mathbf{q}\,|H_{qx}|\,x;\mu\rangle}{(z-E_X+\omega(q)-C_{X\mu}(z+\omega(q)))^2}. \tag{12}$$

By taking thermal average of both sides of the above equation and taking into account Eq.11 we get

IOP Conf. Series: Journal of Physics: Conf. Series **1068** (2018) 012007 doi:10.1088/1742-6596/1068/1/012007

$$\frac{dC_X(z)}{dz} = -\sum_q \left\{ \frac{|g_q|^2 (1 + n(q))}{(z - E_X - \omega(q) - C_X(z - \omega(q)))^2} + \frac{|g_q|^2 n(q)}{(z - E_X + \omega(q) - C_X(z + \omega(q)))^2} \right\}, \tag{13}$$

where

$$C_X(z) = \sum_\mu p_\mu \langle x; \mu | C(z) | x; \mu \rangle \tag{14}$$

with p_μ being the probability to find the reservoir in the state $|\mu\rangle$, and $n(q) = \left[e^{\omega(q)}/k_B T - 1\right]^{-1}$ is mean phonon occupation number at bath temperature T. Here we use natural units ($\hbar = c = k_B = 1$). When temperatures are low enough, this equation can be solved perturbative. At leading order, one can neglect the self-energy function in the denominators on the right-hand part of the equation, and get the leading-order solution

$$C_X^{(2)}(z) = \sum_q \left\{ \frac{|g_q|^2 (1 + n(q))}{E - E_X - \omega(q) + \frac{i\Gamma}{2}} + \frac{|g_q|^2 n(q)}{E - E_X + \omega(q) + \frac{i\Gamma}{2}} \right\}, \tag{15}$$

where $\frac{i\Gamma}{2}$ is added to the denominators on the right-hand part of this equation to account for the relaxation of states. This expression for the self-energy function is exactly the same that has been derived within independent boson model (IBM) at the second Born approximation level [26,27].

At the same time Eq.15 can be solved beyond the perturbation theory. For this purpose, it is convenient to rewrite the equation in the form

$$\frac{dC_X(z)}{dz} = \frac{S_{HR}}{\omega_b^2} \int q^3 e^{\frac{-q^2}{2}} \left\{ \frac{1 + n(q)}{(z - \omega(q) + \frac{i\Gamma}{2} - C_X(z - q))^2} + \frac{n(q)}{(z + \omega(q) + \frac{i\Gamma}{2} - C_X(z + q))^2} \right\} dq, \tag{16}$$

where S_{HR} is dimensionless Huang-Rhys parameter characterizing the exciton-phonon coupling, and $\omega_b = s/l$ is a cut-off frequency (s is the sound velocity and l is the dot size). In Figs.1,2 we show that the self-energy function derived from the nonperturbative solutions of Eq.16 differ profoundly from that obtained at the second-Born perturbation level.

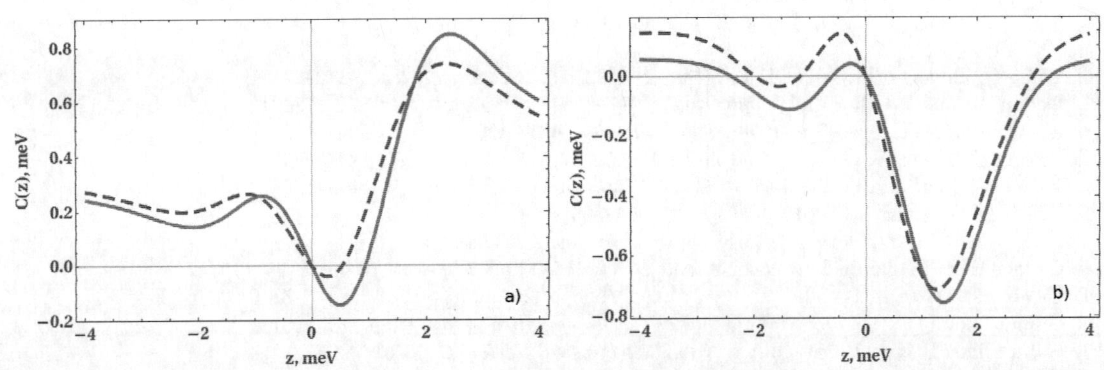

Figure 1. Real (a) and imaginary (b) parts of quantum fluctuation contribution to self-energy function for InAs QDs at $T = 10\,K$, $\Gamma = 0.077\,meV$ and $S_{HR} = 0.2$. The solid lines represent self-energy function calculated within second-Born approximation, dashed lines represent function calculated beyond the perturbation theory

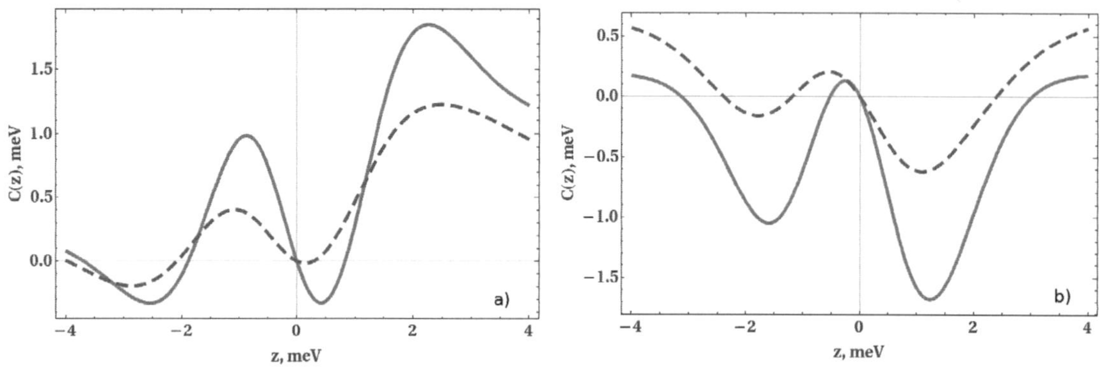

Figure 2. Real (a) and imaginary (b) parts of quantum fluctuation contribution to self-energy function for InAs QDs at $T = 40K$, $\Gamma = 0.077meV$ and $S_{HR} = 0.2$. The solid lines represent self-energy function calculated within second-Born approximation, dashed lines represent function calculated beyond the perturbation theory

4. Conclusion

In conclusion, we have shown that the generalized dynamical equation provides an effective way to solve the problem of the strong interaction between a single QD and phonon reservoir beyond the perturbation theory. In this way, a nonperturbative solution for exciton self-energy function has been found. The results presented in Figs.1,2 show that at temperatures more than 10 K and parameter S_{HR} equal to 0.2 the nonperturbative solution differ profoundly from that obtained in the second-Born approximation [27] to describe the emission from the QD-cavity systems.

References

[1] Woggon U 1997 *Optical properties of semiconductor quantum dots* (Springer)
[2] Song H Z, Hadi M, Zheng Y, Shen B, Zhang L, Ren Z, Gao R and Wang Z M 2017 *Nanoscale research letters* **12** 128
[3] Zhang J, Huo Y, Rastelli A, Zopf M, Hofer B, Chen Y, Ding F and Schmidt O G 2014 *Nano letters* **15** 422–427
[4] Yuan Z, Kardynal B E, Stevenson R M, Shields A J, Lobo C J, Cooper K, Beattie N S, Ritchie D A and Pepper M 2002 *science* **295** 102–105
[5] Yadav A, Bai L, Yang Y, Liu J, Kaushik A, Cheng G J, Jiang L, Chi L and Kang Z 2017 *Nanoscale* **9** 5049–5054
[6] Wang C, Zhuang J P, Grillot F and Chan S C 2016 *Optics express* **24** 29872–29881
[7] Delteil A, Sun Z, Fält S and Imamoğlu A 2017 *Physical review letters* **118** 177401
[8] Loss D and DiVincenzo D P 1998 *Physical Review A* **57** 120
[9] Briegel H J, Dür W, Cirac J I and Zoller P 1998 *Physical Review Letters* **81** 5932
[10] Verzelen O, Ferreira R and Bastard G 2002 *Physical review letters* **88** 146803
[11] Wei H R and Deng F G 2016 *JOSA B* **33** 804–809
[12] Li T, Yang G J and Deng F G 2016 *Physical Review A* **93** 012302
[13] Gauger E M, Benjamin S C, Nazir A and Lovett B W 2008 *Physical Review B* **77** 115322
[14] Calarco T, Datta A, Fedichev P, Pazy E and Zoller P 2003 *Physical Review A* **68** 012310
[15] Gainutdinov R K 1999 *Journal of Physics A: Mathematical and General* **32** 5657
[16] Gainutdinov R K and Mutygullina A A 2002 *Physical Review C* **66** 014006
[17] Gainutdinov R K and Moutygoullina A 1997 *Physics of Atomic Nuclei* **60** 841–847
[18] Gainutdinov R K and Mutygullina A A 2002 *Physics of Atomic Nuclei* **65** 1421–1430
[19] Gainutdinov R K and Mutygullina A 2009 *Physica Scripta* **2009** 014021
[20] Gainutdinov R K 1989 *Journal of Physics A: Mathematical and General* **22** 269
[21] Gainutdinov R K, Mutygullina A A and Scheid W 2002 *Physics Letters A* **306** 1–9
[22] Gainutdinov R K 1983 *Sov. J. Nucl. Phys.* **37** 277–282
[23] Gainutdinov R K 1987 *Sov. J. Nucl. Phys.* **46** 743–749

IOP Conf. Series: Journal of Physics: Conf. Series **1068** (2018) 012007 doi:10.1088/1742-6596/1068/1/012007

[24] Gainutdinov R K 1987 *Sov. J. Nucl. Phys.* **53** 885–892
[25] Gainutdinov R K 1995 *Zh. Eksp. Teor. Fis.* **108** 1600–1613
[26] Tarel G and Savona V 2010 *Physical Review B* **81** 075305
[27] Hughes S, Yao P, Milde F, Knorr A, Dalacu D, Mnaymneh K, Sazonova V, Poole P, Aers G, Lapointe J *et al.* 2011 *Physical Review B* **83** 165313

IOP Conf. Series: Journal of Physics: Conf. Series **1068** (2018) 012008 doi:10.1088/1742-6596/1068/1/012008

Development of technology for state registration signs of the vehicles recognition in conditions of low resolution on the basis of gradient busting and convolutional neural networks

D Z Galimullin[1,2], A I Denisov[2], N V Denisova[3], M E Sibgatullin[4, 5]

[1]Kazan Innovative University, Kazan 420111, Russia
[2]"Veroyatnost", LLC, Kazan 420039, Russia
[3]Kazan State Power Engineering University, Kazan 420066, Russia
[4]Institute of Physics, Kazan Federal University, Kazan 420008, Russia
[5]Institute of Applied Research, Tatarstan Academy of Sciences, Kazan 420111, Russia

E-mail: galimullin_d.z@mail.ru

Abstract. This paper presents the technology of characters on the vehicle license plate recognition. The Viola-Jones method is used for distinguishing the area of the license plate on the basis of pre-trained Haar cascades. Convolutional neural networks analyzing the selected frames are used for recognition. Training procedure of neural networks is carried out through the created synthetic database containing characters with various types of distortion. The efficiency of neural network training was achieved due to the possibility of a sampling formation with the almost unlimited dimension, modeling of distortions which are rarely found in a real sampling due to its limitations, the possibility of forming a mixed sampling including both synthetic and real data. The necessary methods of augmentation and preprocessing for synthetic and real data are implemented, based on the resolution of the video camera. The results obtained in this work can be used for character recognition in conditions of low resolution of the video camera, distortions and lack of initial data.

The first successful applications of neural networks of deep training belong to the beginning of the 90s of the last century. The AT&T Bell Labs under the guidance of Yann LeCun developed a system of automatic recognition of handwritten figures. However, due to the complexity of convolutional neural networks training, the interest of researchers over an extended period of time has been concentrated on other approaches to machine training. Progress in the hardware and research fields has led to a renewal of interest in this method on the part of the scientific community. Particularly strong interest was witnessed after the 2010s, when such networks showed better results at recognizing images, speech, etc. At present, the systems based on the use of neural networks of in-depth training show the best results in image recognition on image test bases, such as MNIST, CIFAR-10, IMAGENET [1-3].

The paper presents the technology for the symbols of the Russian Federation vehicle license plates recognizing by the optical video flow on the basis of the gradient boosting and convolutional neural networks in the gray-scale mode and low-resolution conditions. The mechanism for synthetic training samples generating was developed and implemented, as well as a separate database of marked data relevant to the real conditions of data sampling and occurring distortions. The process of optimization, in comparison with [4], led to the network structure simplification, additional procedures inclusion, increase of the training efficiency, reduction of the neurons number in the intermediate layers without significant loss in accuracy.

Generation of synthetic marked out data

The first stage in the recognition technology construction is a preliminary search for the vehicle license number, which consists in finding the area of the image containing the number. The Viola-Jones method [5-7] was used to indicate a fragment with a number plate.

It was necessary to prepare a marked out database for training the neural network. The task of preparing a high quality training sampling was one of the most important ones; besides, it was essential that the prepared data be relevant to the real data obtained during a video stream processing. To that effect, we carried out a detailed analysis of the obtained numbers database after the Haar cascades development, and detected the following distortions appearing in the frame (Fig. 1):

1. Distortions of the symbols, caused by their "blurring" due to the vehicle movement, weather conditions, considerable distance to the camera while fixing the object of interest.
2. Distortions of the background, caused by noise due to weather conditions, take-off time, camera quality, or contamination of the license plate area.
3. Distortions of symbols due to the fixation method or angle of the video camera, the form of the vehicle and method of the number fixing. Such distortions lead to geometric distortions in the symbol shape or to a change in its slope angle compared to a standard one.

Fig.1. Distortions of symbols and a background appearing on the vehicle license number

Since it would be really difficult to create a real database of marked out numbers with all possible distortions included, due to the limited access to such data, it was decided to create a mechanism for generating artificial license plate numbers based on a real prototype. We should note the advantages of the synthetic training sampling. It is an opportunity to form a sampling of almost unlimited dimension, simulate those distortions that are rarely found in a real sampling, and use both synthetic and real data.

For creation of synthetic license plate numbers and simulation of distortions, we developed an algorithm [4], which served the basis for synthetic numbers generation. The example of generated images is shown in Fig.2.

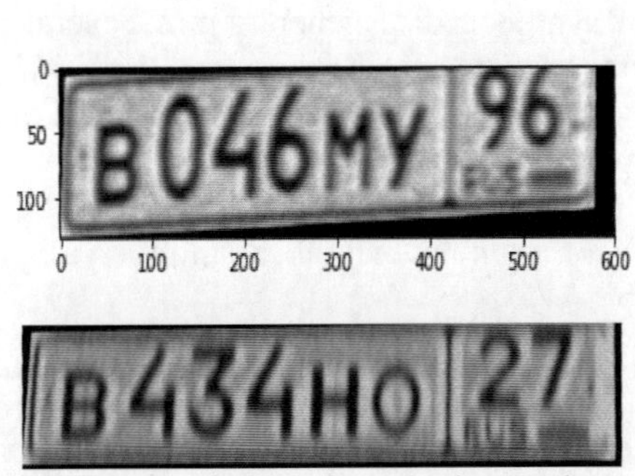

Fig.2. The example of synthetic numbers

Cutting out fragments containing symbols with random distortions inserted with the help of a specially placed window was the next stage. Further, the image data was recorded into the database with the appropriate labels (Fig. 3).

IOP Conf. Series: Journal of Physics: Conf. Series **1068** (2018) 012008 doi:10.1088/1742-6596/1068/1/012008

Fig. 3 Symbols generated for the database

Creating a training dataset

The reference [4] presents a synthesized training sample and conducted network training. The neural network showed about 99% accuracy in the test data. However, the initial accuracy did not exceed 60% when working with the real data. The carried out analysis showed that the neural network is susceptible to the quality and dimension of the input data displayed in the sliding window. If a degree of the camera resolution is less than the resolution of the images used for training, then the recognition quality is low, especially with the occurring distortions of any kind. Accordingly, it was necessary to optimize the generated synthetic data adapting them for the corresponding characteristics of the video stream images to process the real data received from the video camera. For quality improvement, it was necessary to generate samples of a larger volume with a sufficiently high diversity. It guaranteed greater stochasticity for the stable and accurate operation of the network.

In order to understand the particular distortions to be used, some of the real data was applied for validation in the synthetic-real ratio as 5 to 1. To improve the quality, synthetic data was mixed with real data from a database that was freely available on the Internet in proportions of about 5 to 1. As a result, a part of the real data comprised about 40,000 samples, including 20,000 augmented ones. The volume of the synthetic sampling comprised 250,000 samples, including 125,000 augmented ones.

The following procedures were carried out for the input data. The dimension was reduced to 16x16 pixels. This allowed speeding up the network operation and increasing the resistance to noise distortion. A unified image database was developed, which included several samplings of synthetic images obtained at different parameters mixed with the real data. The size of the final dataset made 300,000 samples.

The size of the training and validation sampling was in the ratio 4:1. Synthetic images were generated by groups with different resolutions and then were reduced to the original dimension of 16x16 pixels. In the process of synthetic data generating, a general approach to their creation was developed and the following number of transformations were used (Fig. 4):

1) Application of Gaussian noise in various proportions, application of Gaussian filters for blurring of symbols.
2) Applying distortions to basic images simulating fragments of symbols.
3) Zooming by adjusting the area of a symbol capture, resulting in a random shifts of the symbol along the X and Y axes.
4) Random turns to an angle up to 5 degrees.
5) Affine transformation with random choice of parameters in a certain range.
6) Conversion of perspective with random selection of parameters in a certain range.

The synthetic data were augmented at the next stage, using the Keras library functional and included the following procedures [8]:

1) Turns to an angle up to 5 degrees.
2) Compressing and stretching the data along the X and Y axes by 10% of the original dimensions.
3) Random shifts along the X and Y axes by 10% of the original dimensions.

4) Correlation between the neighboring colors of the pixels can adversely affect the neural network. Therefore, the ZCA Whitening procedure [9] based on the principal component method was used to eliminate correlations.

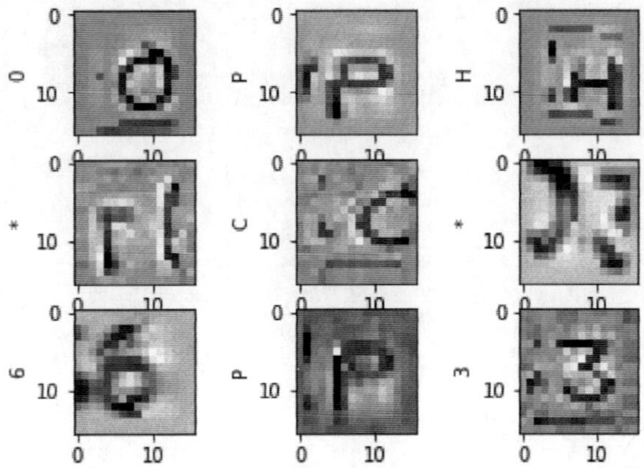

Fig. 4 Synthetic symbols augmentation example. Noise fragments are marked out with *.

Further, augmentation was carried out with the help of the same operations for the data from the real database, filmed by Russian license plate numbers and using the Keras library. The significant shortcomings of real data are due to the fact that the sizes of many symbols are fixed and occupy a certain area of the image; they have a uniform background and a fixed form, since the sampling is made under certain specific conditions. Training without augmentation here is ineffective, as the real images can be significantly distorted due to other conditions of the sampling, camera resolution, external influences. An example of symbols augmentation is shown in Fig.5 After this procedure, the shape of the symbols, their position in the area, the color correlation on the background were significantly altered.

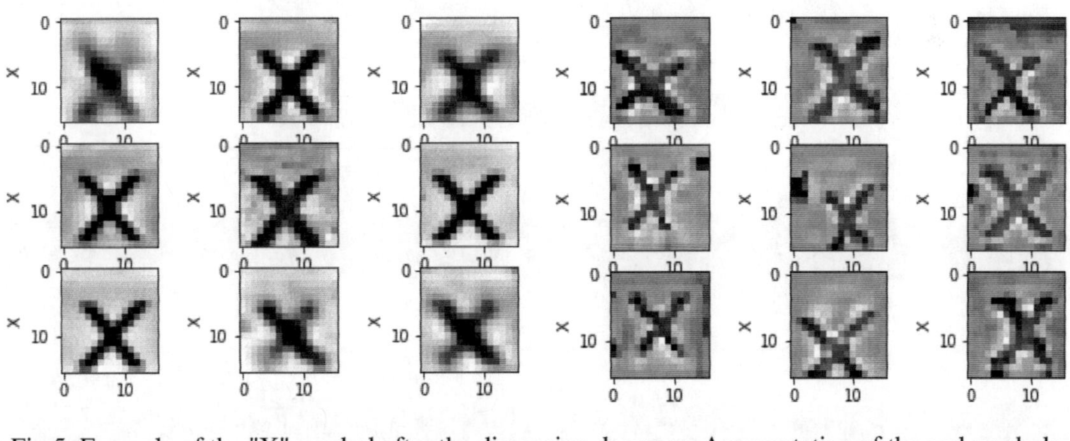

Fig.5. Example of the "X" symbol after the dimension decrease. Augmentation of the real symbols.

Recognition module optimization

The next stage consisted in building an algorithm which allows to "collect" together all the detected symbols into a license plate, while cutting off false signals. Judging from the size of the fragment [8] the center was determined and the dimensions were set for the window. Further, when the image passed the window, the neural network determined the symbol with the maximum probability, its value and probability being recorded in the corresponding arrays (Fig. 6).

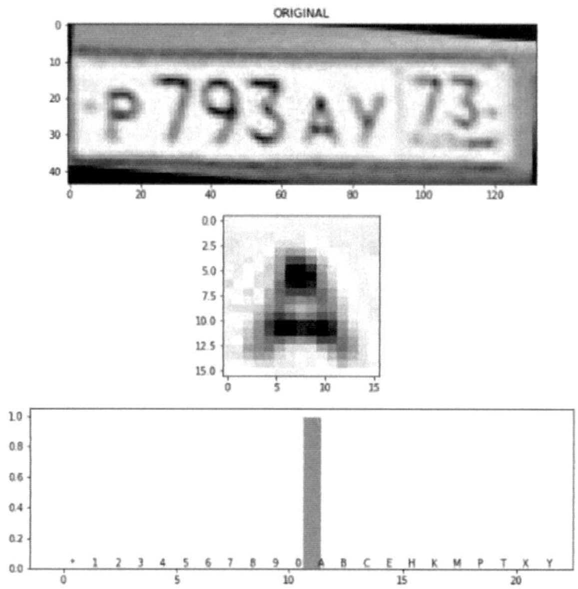

Fig. 6 License plate number, symbol in the sliding window and its probability

The shift was equaled to 1 pixel. Further, the probability evaluation was carried out. For each of the detected symbols, the number of its occurrences (along the X axis) and the probability for each case are shown. The Fig. 7 shows that the probabilities greater than 90% have the correct values.

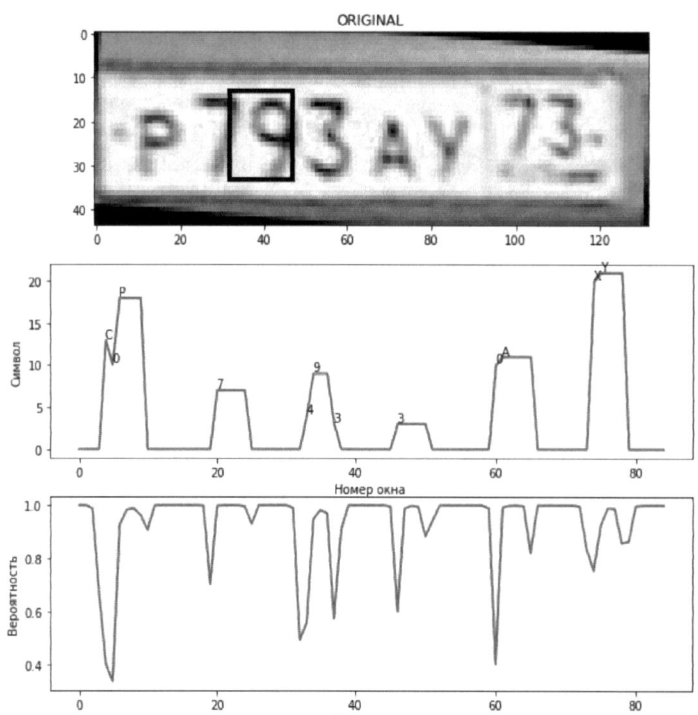

Fig.7. The initial image, symbols detected at scanning by the window and their probabilities.

Further, the sum of all probabilities was calculated for each registered symbol, except for the zero class marked out with * in the Fig. 6, which is the noise. All elements of the noise class were cut off. The first six, which have the maximum values, form a license plate. Together with the sum of probabilities value, they were recorded in a certain list (Fig. 8), which was indexed.

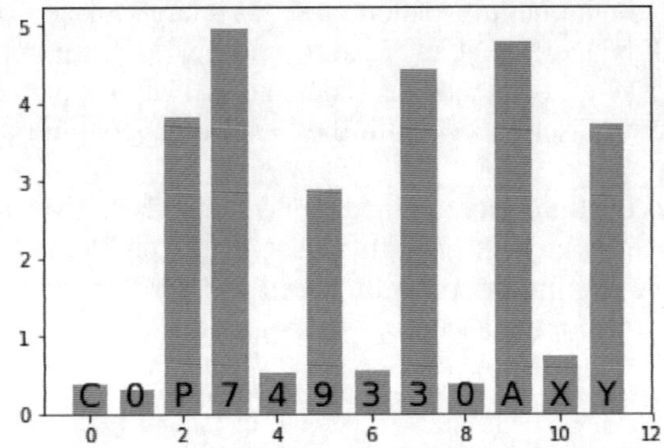

Fig. 8 Sum of probabilities for all detected symbols

Such kind of procedure allows selecting the most correct characters and numbers in the probabilistic sense. At the output after restoring the order of the coming symbols, we get the collected license plate number ['P', '7', '9', '3', 'A', 'Y'], the vector of probability sums for each symbol [3.86, 4.99, 2.9 , 4.46, 4.81, 3.76], and the result of these values addition is 24.78.

The multi-window approach was proposed to reduce errors in recognition, which included the following procedure. For the purpose of the recognition authenticity increase, an additional window shift along the Y axis relative to the original position was carried out, and the characters were scanned and recognized for each window position. The shift vector was determined by the values [-6, -4, -2, 0, 2, 4, 6].

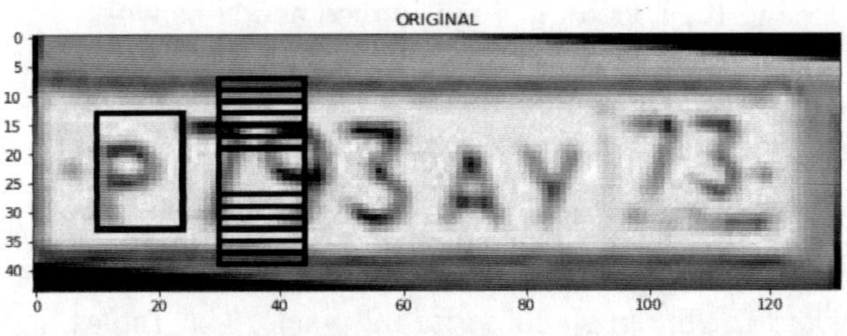

Fig. 9 The initial image and different positions of windows.
The window with optimal position is indicated separately.

Figure 9 shows all possible positions of the window. We calculated the sum of the elements of the vector for each of the variants with components equal to the sums of probabilities in respect of each symbol. The maximum value indicates the optimal window position (highlighted by a rectangle), which is selected for further data processing. Later, at data processing for speeding up the analysis, we limited ourselves to a vector of shifts with the values [-6, -4, -2, 0, 2, 4, 6], since for 95% of cases this would be sufficient for correct recognition.

Optimization of the structure and speed of the neural network

For maximum performance, the network structure was similar to Lenet structure [2] with some changes. Its effectiveness was shown in the reference [4]. The topology of the network used is shown in Fig. 10.

Since speed was required alongside with a high recognition quality, so its structure had to be simplified as much as possible on the basis of general considerations and the work done at the previous stage. Obviously, the effective operation required the use of at least two convolutional layers, but at the same time the reduction of the number of filters without the decrease in recognition quality. It is necessary to place a decrease of dimension layer between the first and the second convolutional layers (max_pooling2d) with a minimal dimension mask, as a larger mask dimensions for images with small image dimensions shall be inefficient and lead to a strong decrease in the data dimension before the second layer. The classical ReLu-function served as an activation function. The dropout inclusion with a certain number of neuron disconnections in the Lenet-type structure made it possible to reduce the number of iterations [10].

It should be noted that this configuration, trained on synthetic data, gives an accuracy of at least 98-99% on the validation dataset, which is enough to solve our problem. In addition, the minimum number of layers provided the maximum operation speed and allowed estimating the influence of the number of convolutional neurons for the accuracy of recognition.

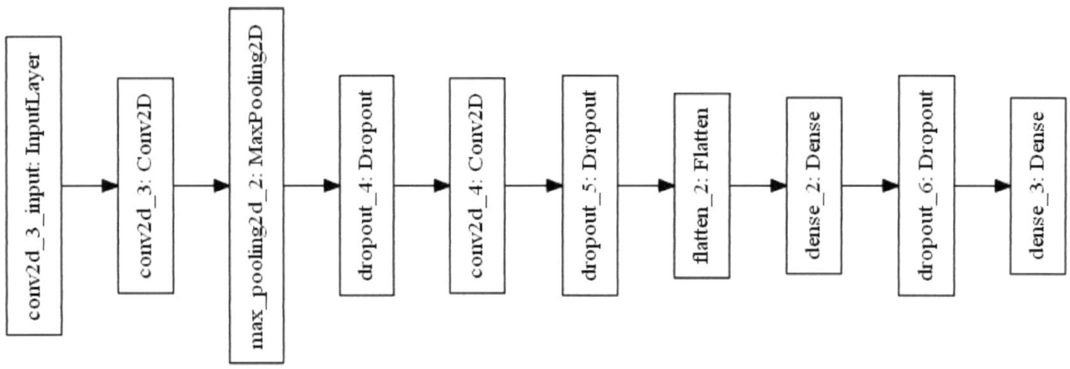

Fig.10. Topology of the applied neural network

The experimental estimation of operation speed and recognition accuracy for different ratios of neurons in convolutional and fully connected layers was carried out for optimizing the network structure. The used configurations, the operation speed with a single pass and multi-window mode and the recognition accuracy are shown in Table 1. Multi-window mode presents a multiple pass by a sliding window with a different height offset of the image for reduction of the recognition error. The number of shifts was equal to 7, the pitch comprised 2 pixels. The Table1 shows that the values for one pass and multi-window mode are averaged values over 40 iterations.

Table 1. Applied topologies, their operation speed and accuracy of recognition

Number of the neural network structure	1	2	3	4	5	6	7	8	9
Convolutional layer No1, number of masks	4	4	8	8	16	16	16	16	32
Convolutional layer No2, number of masks	4	4	4	8	8	8	16	16	64
Fully connected layer, Number of neurons	256	512	512	512	256	512	512	1024	1024
Time, ms (One pass)	36	38	41	44	46	48	52	65	180

Time, ms (Multi-window mode)	255	260	280	293	313	325	355	450	1300
Recognition accuracy at the validation dataset	97.3	98.1	98.2	98.7	98.8	99	99.2	99.3	99.5

Conclusion

The research work has optimized the mechanism for creating the synthetic training sampling based on synthetic license plate numbers. A database of marked out license plate number symbols (according to GOST R 50577-93), relevant to the real conditions of data sampling, low camera resolution (16x16 pixels per area with a symbol) and the resulting distortions is developed. The basic configuration of the network for working with the real data is optimized in terms of speed and quality of recognition, and the training mechanism based on a synthetic learning sampling is implemented. Necessary methods of augmentation and preprocessing for synthetic and real data are implemented, based on the video camera resolution.

An algorithm for recognizing license plate numbers applying deep neural networks directly by the video stream is proposed, including a license plate number detector based on the Haar cascades, the multi-window approach for correct detection of license plate number and region symbols. The permissible license plate angle shall not exceed 10°. The recognition probability from 90% to 99% depends on the conditions of visibility and quality of license plates.

The results can be used for low-budget solutions in symbols recognition in conditions of low resolution of the camera, distortions and lack of the initial data. The approach in respect of the synthetic data use can be generalized for the other classes of objects.

The work was fulfilled under the support of the Bortnik Foundation (No. 1457GS1 / 22736).

References

[1] Sermanet P, Chintala S, LeCun Y 2012 *Convolutional Neural Networks Applied to House Numbers Digit Classification* International Conference on Pattern Recognition (ICPR 2012) pp.3288-3291

[2] Ian Goodfellow J, Bulatov Y, Julian I, Arnoud S, Shet V 2014 Multi-digit Number Recognition from Street View Imagery using Deep Convolutional Neural Networks (arXiv:1312.6082v4)

[3] Jaderberg M, Simonyan K, Vedaldi A, Zisserman A 2014 Reading Text in the Wild with Convolutional Neural Networks (arXiv:1412.1842v1)

[4] Galimullin D Z, Denisov A I, Illarionov S Yu 2017 Convolutional neural networks training effectiveness estimation basing on the synthetic dataset for recognition of the state registration license plate number symbols *Postulate* (http://e postulat.ru/index.php/Postulat/article/view/653/676)

[5] Viola P, Jones M J 2004 Robust real-time face detection *International Journal of Computer Vision* **57**(2) 137-154

[6] Wu X, Kumar V, Ross Quinlan J at all 2008 Top 10 algorithms in data mining *Knowl. Inf. Syst.* pp.1-37

[7] Chaudhari M, Sondur S, Vanjare G 2015 A review on Face Detection and study of Viola Jones method *International Journal of Computer Trends and Technology (IJCTT)* **25**(1) pp.54-61

[8] https://keras.io/preprocessing/image/

[9] Kessy A, Lewin A, Strimmer K 2016 Optimal whitening and decorrelation (arXiv:1512.00809)

[10] Srivastava N, Hinton G, Krizhevsky A, Sutskever I, Salakhutdinov R 2014 Dropout: A simple way to prevent neural networks from overfitting *Journal of Machine Learning Research* pp.1929-1958.

FTIR spectroscopic study of conformational equilibria of substituted ethane in polyethylene glycols

D I Kamalova, S N Naumova

Kazan Federal University, Kazan 420008, Russian Federation

E-mail: dina.kamalova@kpfu.ru

Abstract. In this paper, we considered the conformational behaviour of 1,2-dichloroethane in polyethylene glycols with molecular weight equaled to 200, 400 and 600. We determined the enthalpy differences of conformations of this compound in polyethylene glycols by FTIR spectra. The obtained values were compared with that for high molecular weight polyethylene glycol PEG 10,000.

1. Introduction

Polyethylene glycols with low molecular weight are widely applied in the chemical, pharmaceutical, food industry as plasticizers [1-3]. Polyethylene glycol with average molecular weight is used as a component of solid polymer electrolyte films for conductors of ions in lithium-polymeric elements to product accumulators [4]. Electroconductive polymers with internal conductivity are the molecular systems which are similar to well-known conductors on the conductivity mechanism. Solid polymeric electrolytes draw special attention in connection with broad application as electrochemical systems, such as fuel elements, electrolyzers, lithium accumulators [5, 6]. Solid polymeric electrolytes are solutions of salts in a polymeric matrix as a macromolecular solvent. Polyethylene glycols PEG 200, PEG 400, PEG 600 are used as plasticizers in solid electrolyte films for cell model [4]. As shown in [4], PEG 400 is more effective than others in enhancing the conductive performances of these solid polymer electrolytes. When solid polymeric electrolyte creating for fuel elements, it is necessary to take into consideration different physical factors which influence on electric conductivity, including thermal mobility of fragments of a polymeric chain and formation of channels for transport of ions. The purpose of this work is to study conformational behavior of 1,2-dichloroethane (DCE) in low molecular weight polyethylene glycols and to determine the difference of the enthalpy of conformations by FTIR spectroscopy.

In liquid and solutions, this conformationally-inhomogeneous compound is in two conformations (*trans* and *gauche*), between which a dynamic equilibrium may be established [7]. Two analytical bands relating to the *trans* and *gauche* conformations are identified in the IR spectrum. For their integral optical density D, the Beer–Lambert–Bouguer law is implemented:

$$D_{trans} = \alpha_{trans} l c_{trans}, \qquad D_{gauche} = \alpha_{gauche} l c_{gauche},$$

where α_{trans} and α_{gauche} are the integrated absorption coefficients of the respective bands, l is the thickness of the absorbing layer, c_{trans} and c_{gauche} are the concentrations of *trans* and *gauche* conformations. The concentration ratio is determined by equilibrium constant K, which depends on the temperature:

$$K = c_{trans}/c_{gauche} = \exp(-\Delta G_0/RT),$$

where ΔG_0 is the difference in free energy of conformations, which can be represented as the difference between the enthalpy and entropy terms:

$$\Delta G_0 = \Delta H_0 - T\Delta S_0,$$

where ΔH_0 and ΔS_0 are the difference of the enthalpy and difference of entropy of conformations, respectively. For the ratio of the optical densities of analytical absorption bands of *trans* and *gauche* conformations, we have

$$\ln(D_{trans}/D_{gauche}) = \ln(\alpha_{trans}/\alpha_{gauche}) - \Delta H_0/RT + \Delta S_0/R + \text{const.} \quad (1)$$

The values ΔH_0 and ΔS_0 are thermodynamic parameters of conformational equilibrium. $\Delta H_0/R$ is defined as a slope of the linear dependence of $\ln(D_{trans}/D_{gauche})$ upon $1/T$ at the temperatures which are higher than freezing temperature of conformational transitions.

2. Experimental

Polyethylene glycols (PEG 200, PEG 400, PEG 600, PEG 10,000) were used as received. These chemicals were purchased from Sigma-Aldrich. DCE (Acros organics) was used as conformationally-inhomogeneous compound. Structural formula of monomer units of PEG is

$$H - [-O - CH_2 - CH_2 -]_n - OH.$$

PEG 200 and PEG 400 are liquid at room temperature while PEG 600 is waxy solid moist. Sample for DCE in PEG 600 was prepared by heating and then solving DCE in PEG 600. The obtained solution was spread between KBr plates. The DCE concentrations were 3-5 wt %. The concentration and the thickness of the absorbing layer were calculated by IR spectra. Films of PEG 10,000 were prepared by dissolving in chloroform.

The analytical conformation-sensitive absorption bands were at 654 (*gauche*) and 710 (*trans*) cm^{-1} for DCE [8]. Analytical absorption bands did not overlap absorption bands of PEG. Fourier-transform infrared spectra were recorded with a Perkin Elmer FTIR spectrometer Frontier. We used the cryostat Specac cooled by liquid nitrogen.

3. Results and discussion

FTIR spectra of studied PEGs at room temperatures are represented in Fig. 1. For comparison the FTIR spectrum of high weight molecular PEG 10,000 is also given in Fig. 1. As seen, the all spectra are similar, with the exception that in the spectra of PEG 200, PEG 400 and PEG 600, there is the broad band arising from hydrogen-bonded –OH at ~3350 cm^{-1} while in the spectrum of PEG 10,000 this absorption band is absent. There is not the hydrogen-bonded structure in PEG 10,000 with high degree of crystallinity.

FTIR spectra of systems DCE in PEGs have been obtained at different temperatures in the range of 300-200 K. As an example, the spectrum of DCE in PEG 200 is represented in Fig. 2. The change of intensities of conformationally-sensitive analytical absorption bands (*gauche* and *trans*) of the DCE takes place when temperature changes in the temperature range under study. Similar changes have been observed for other systems DCE in PEGs. Crystallization of DCE solutions in low molecular PEGs did not occur at the temperature range under investigation.

The value ΔH_0 in Eq. (1) is the difference of the enthalpy of conformations which may be estimated by the slope of the dependence $\ln(D_{trans}/D_{gauche})$ upon $1/T$. The magnitudes ΔH_0 for DCE in PEGs are listed in Table. For comparison, ΔH_0 for DCE in PEG 10,000 are also given in Table. The obtained differences of the enthalpy of conformations of DCE in low molecular PEGs as media reflect the contributions of the intermolecular interactions. Effect of medium on conformational equilibrium of molecules is the result of complex of various interactions between the molecules of DCE and the environment. If to consider that the dominating contribution to the difference of enthalpies of conformations is made by electrostatic interactions [7], then this is confirmed by linear correlation between ΔH_0 and the Onsager function $\sigma = (\varepsilon-1)/(2\varepsilon+1)$ where ε is the dielectric permittivity of medium [9].

IOP Conf. Series: Journal of Physics: Conf. Series **1068** (2018) 012009 doi:10.1088/1742-6596/1068/1/012009

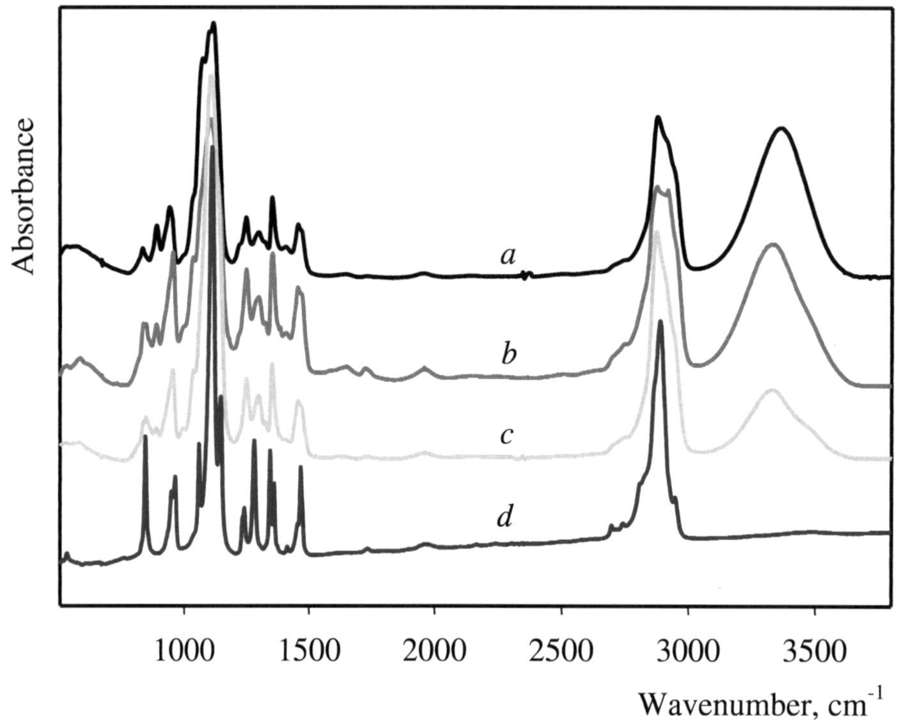

Figure 1. FTIR spectra of PEG 200 (*a*), PEG 400 (*b*), PEG 600 (*c*) and PEG 10,000 (*d*) at room temperature

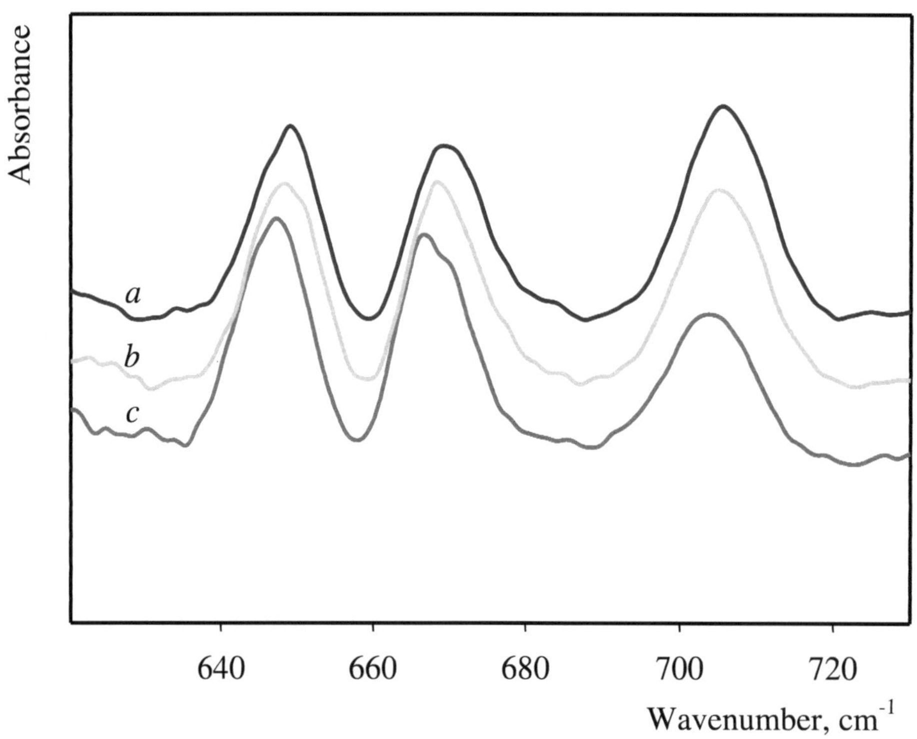

Figure 2. FTIR spectra of DCE in PEG 200 at different temperatures: 273 (*a*), 243 (*b*) and 193 (*c*) K

IOP Conf. Series: Journal of Physics: Conf. Series **1068** (2018) 012009 doi:10.1088/1742-6596/1068/1/012009

The obtained ΔH_0 values for DCE in low weight molecular PEGs (σ = 0,22) satisfy this correlation. As seen from Table, the magnitudes of ΔH_0 for DCE in low weight molecular PEG 200, PEG 400 and PEG 600 are approximately equal to each other, that is, ΔH_0 does not depend on weight of low molecular PEG. As to DCE in PEG 10,000, the value ΔH_0 is less. It is explained by the difference of cavity energy in liquid media and polymeric matrix. It means that cavity interactions contribution to the difference of enthalpies of conformations is less in PEG 10,000 as compared with low weight PEGs.

Table. Enthalpy differences of conformations ΔH_0 (kcal/mole) for DCE in PEGs with different molecular weight

PEG	PEG 200	PEG 400	PEG 600	PEG 10,000
ΔH_0	0,84	0,86	0,87	0,63

*) The errors in the ΔH_0 evaluation do not exceed 5% of the determined magnitudes

4. Conclusion

We have considered the conformational behaviour of 1,2-dichloroethane in polyethylene glycols with molecular weight equaled to 200, 400 and 600. We have shown that the enthalpy differences of conformations of 1,2-dichloroethane in polyethylene glycols are approximately identical. The cavity contribution to enthalpy difference of conformations of 1,2-dichloroethane in PEG10,000 is less than that PEG 200, 400, 600 in consequence of the rigid polymeric matrix of PEG10,000.

References

[1] Vito Di Noto 2002 *J Phys Chem B* **106** 11139-11154
[2] Ji Chen, Scott K Spear, Jonathan G Huddleston and Robin Rogers 2005 *Green Chem* **7** 64-82
[3] Pin Jiang Li, Ji Huai Wu, Miao Liang Huang, San Cun Hao, Zhang Lan, Qinghua Li, Shijie Kang 2007 *Electrochimica Acta* **53** 903–908
[4] Li Yan and Jun Wang 2011 *Desalination* **281** 455-461
[5] Amrtha Bhide, K Hariharan 2007 *European Polymer Journal* **43** 4253-4270
[6] M Jaipal Reddy, J Siva Kumar, U V Subba Rao, Peter P Chu 2006 *Solid State Ionics* **177** 253-256
[7] Abraham R J, Bretschneider E 1974 in *Internal Rotation in Molecules*, edited by W J Orville-Thomas (Wiley, London)
[8] Sverdlov M M, Kovner M A 1970 *Kolebatelnie spektri mnogoatomnikh molekul* (Moscow: Nauka)
[9] Minoru Kato, Isao Abe, Yoshihiro Taniguchi 1999 *J Phys Chem* **110** 11982-11986

IOP Conf. Series: Journal of Physics: Conf. Series **1068** (2018) 012010 doi:10.1088/1742-6596/1068/1/012010

Modeling transmission light in photonic band gap of one-dimensional photonic-plasmonic crystals with buffer layer

A V Koryukin[1,2], A A Akhmadeev[1,2], A F Sitdikova [1] and M Kh Salakhov[1,2]

[1] Kazan Federal University, Institute of Physics, 420008, Kremlevskaya St, 18, Kazan, Russian Federation

[2] Tatarstan Academy of Sciences, Institute of Applied Research, 420111, Bauman St, 20, Kazan, Russian Federation

E-mail: akoryukin@gmail.com

Abstract. The optical transmission wasn't researched completely for the hybrid plasmonic-photonic crystal with buffer dielectric layer between photonic crystal (PC) and layer of gold. In this article we consider the dependence of wavelength of the Tamm plasmon on refraction index of the buffer layer. Wavelength of the transmission peak in the photonic band gap increases at the increasing refractive index of buffer layer between metal and PC. Intensity of the transmission peak in the photonic band gap decreases at the increasing refractive index of buffer layer in metal-PC interface. So we can change distribution of energy inside hybrid photonic-plasmonic mode. This results have a promise application in developing lasers and sensors.

1. Introduction

Every year the interest in plasmonics is growing. Surface plasmons are of interest because of their potential use in chemistry and medicine [1–3], optical computers [4] and photonics [5]. Tamm plasmons (TP) are used in new types of lasers [6] and highly sensitive sensors [7]. Plasmons are collective oscillations of the electron density of a metal [8]. Surface plasmons appear at the metal-dielectric interface, where air is usually used as a dielectric [9]. The combination of photonic crystals and plasmon material layer provides the possibility of exciting a new type of plasmons - Tamm plasmon [10]. Their existence is connected with the influence of dispersion properties of a metal and a photonic crystal (PC) on each other. PC are materials with spatial periodicity in their dielectric function (or, equivalently, a periodical index of refraction) [11]. Such periodicity yields a strong modification of the electromagnetic field that manifests itself in new quantum electrodynamical effects [12–14]. Unlike ordinary surface plasmons, TP can be excited directly, without additional devices, for example prisms or gratings, and they can exist in both polarizations [10, 15]. Hybrid modes are also used in apertureless near-field optical microscopy [16], while the use of conventional surface modes for these purposes requires optimization of the excitation geometry [17]. As a result, an extraordinary transmission is obtained in the region of the forbidden band of the PC (the wavelength range at which the wave can not propagate through the crystal [11,18] and is reflected from it [19]). Such structures in the literature are called Hybrid photonic-plasmonic crystals (HPPC) [20–27]. As is well known, the wavelength of a Tamm plasmon is related to the Bragg wavelength and, correspondingly, to the ratio of the refractive indices of the PC [10]. Within one forbidden band, it is possible to change the wavelength of the transmission peak by varying the order of the PC layers or introducing an additional buffer layer. This is in demand when working with certain materials or on the condition that the wavelength of the TP slightly differs from the Bragg wavelength. The HPPC with the buffer layer have been studied in several articles [25, 28, 29], but the advantage of introducing an additional layer has not been fully investigated. In this paper we investigated the dependence of the wavelength of a Tamm plasmon from the refractive index. As a sample, a hybrid photonic-plasmonic crystal was selected based on a one-dimensional PC.

2. Results of modeling

To study the dependence of the wavelength of a Tamm plasmon of a hybrid photonic-plasmonic crystal, materials with refractive indices of 1.46 and 1.95 and with layer thicknesses of 145 and 110 nm, respectively, were selected. The photonic crystal consisted of six pairs of layers. The thickness of the gold layer was 30 nm. The geometry and optical parameters of the structure are taken from [8]. To simulate the transmission spectra of these structures, Maxwell's equations were solved numerically using the finite-difference time-domain method. The boundary conditions were chosen in the form of perfect matched layers. The modeling scheme is shown in Figure 1.

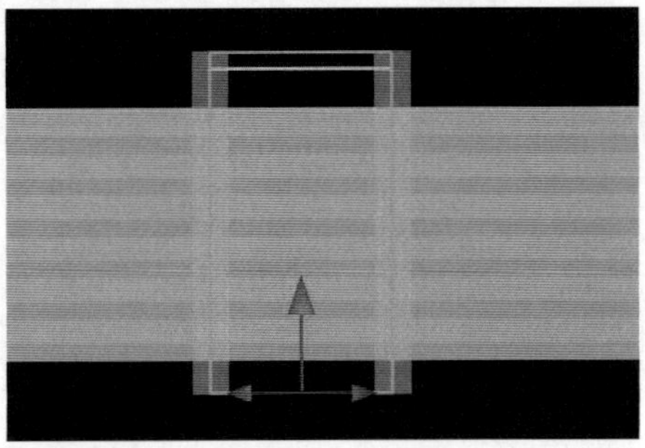

Figure 1. The scheme of modeling. An upward arrow indicates the propagation of light.

The transmittance spectra of a basic one-dimensional photonic crystal and a photonic crystal with a gold layer with the above parameters are shown in Fig. 2. In this figure you can see a characteristic wide transparent zone of a one-dimensional PC and its change in the presence of a gold layer. The transmission intensity in the near infrared region of the spectrum ($\lambda > 750$ nm) decreases significantly down to zero. But there remains a peak associated with the excitation of Tamm plasmon resonance.

A one-dimensional hybrid photonic-plasmonic crystal (HPPC) was modeled with an additional buffer layer, the layer thickness was 80 nm. The refractive index of the buffer layer was 1.45. The spectrum of such a photonic crystal is also shown in Fig. 2. The presence of a buffer layer leads to a shift in the transmission peak to the long-wavelength region and an increase in its intensity by about one and a half times In the future, the aforementioned parameters of a photonic crystal with a buffer and gold layer were used for the simulation. The refractive index of the buffer layer varied from 1.45 to 1.95 in steps of 0.1.

Figure 3 shows the intensity versus the refractive index of the buffer layer. When the refractive index of the buffer layer changes, the intensity first increases, then decreases. This suggests that

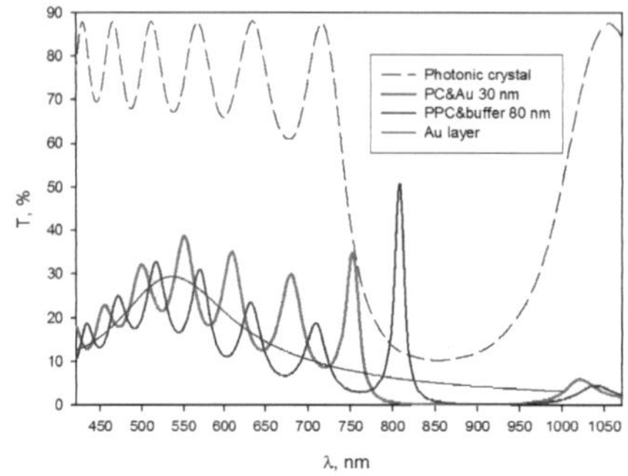

Figure 2. Transmission spectra obtained at an angle of 0: the dotted line is a photonic crystal, the thick line is a PC with a gold film of 30 nm, a thin line is a HPPC with a buffer layer of 80 nm, a dotted line is a thin gold film.

the Tamm mode depends on the refractive index of the buffer layer and there are resonance conditions.

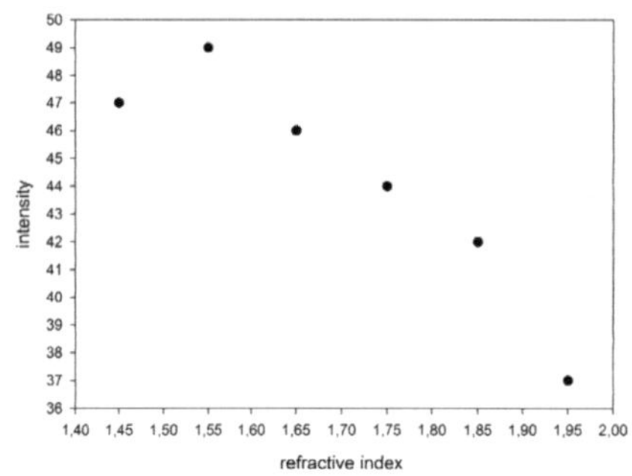

Figure 3. Dependence of intensity on the refractive index of the buffer layer.

Figure 4 shows a plot of wavelength versus the refractive index. The change in the refractive index of the buffer layer leads to a shift in the transmission peak to the long-wavelength region.

Since this transmission peak is associated with plasmon resonance in gold, its displacement means the excitation and propagation of the hybrid photonic-plasmonic crystal mode at the boundary of the buffer layer with gold. This allows us to conclude that the wavelength of the hybrid mode of the Tamm plasmonic depends on the refractive index of the buffer layer. By varying the refractive index of the buffer layer, it is possible to control the energy distribution within the hybrid mode.

3. Conclusion

Based on the simulation results, it was shown that the wavelength of the Tamm plasma depends on the refractive index of the buffer layer. It is also revealed that the intensity of the peak depends on the refractive index of the buffer layer. Thus, by varying the refractive index and the thickness of the buffer layer, it is possible to vary the wavelength of the Tamm plasma within its bound forbidden band. Such an effect of the additional layer can be used in lasers and sensors based on hybrid photon-plasmon crystals.

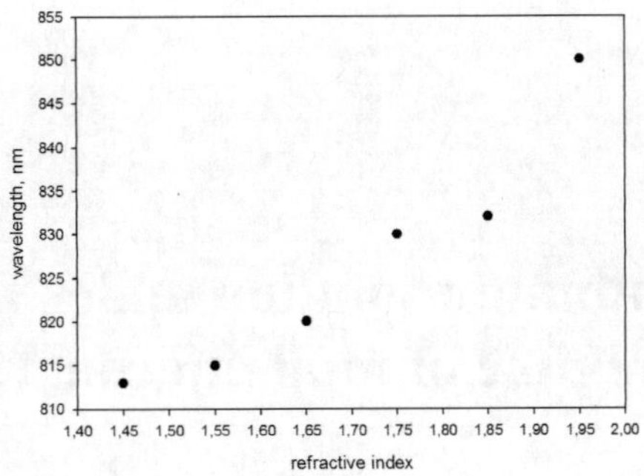

Figure 4. Dependence of wavelength on the refractive index of the buffer layer.

References

[1] Barnes W, Dereux A, Ebbesen T 2003 *Nature* **424** 824-830
[2] Homola J, Ye S, Gauglitz G 1999 *Sens. Actuators B.* **54** 3-15
[3] Kneipp K, Kneipp H, Itzkan I, Dasari R, Feld M *J. Phys.: Condens. Matter.* 2002 **14** 597-624
[4] Krasavin A, Zayats A, Zheludev N 2005 *J. Opt. A, Pure Appl. Opt.* **7** 85-89
[5] Zayats A, Smolyaninov I 2003 *J. Opt. A, Pure Appl. Opt.* **5** 16-50
[6] Symonds C, Lheureux G, Hugonin J P, Greffet J J, Laverdant J, Brucoli G, Lemaitre A, Senellart P and Bellessa J 2013 *Nano Lett.* **13** 31793184
[7] Zhang W L, Wang F, Rao Y J, Jiang Y 2014 *Optic Express.* **22** 14524-14529
[8] Novotny L and Hecht B 2006 *Principles of nano-optics* (New York: Cambridge Univrsity Press) p 539
[9] Maier S 2007 *Plasmonics: Fundamentals and Applications* (Springer Science+Business Media LLC) p 224
[10] Kaliteevski M, Iorsh I, Brand S, Abram R A, Chamberlain J M, Kavokin A V and Shelykh I A 2007 *Phys. Rev.* B **76** 165415
[11] Joannopoulos J D, Johnson S G, Winn J N and Meade R D 2008 *Photonic crystals: molding the flow of light* (Princeton: Princeton University Press) p 286
[12] Gainutdinov R Kh, Khamadeev M A and Salakhov M Kh 2012 *Phys. Rev.* A **85** 053836
[13] Gainutdinov R Kh, Khamadeev M A and Salakhov M Kh 2013 *J. Phys.: Conf. Ser.* **478** 012017
[14] Akhmadeev A A, Gainutdinov R Kh, Hermann G, Khamadeev M A, Steryakov O V and Salakhov M Kh 2015 *J. Phys.: Conf. Ser.* **613** 012005
[15] Sasin M E, Seisyan R P, Kaliteevski M A, Brand S, Abram R A, Chamberlain J M, Iorsh I, Shelykh I A, Egorov A Yu, Vasilev A P, Mikhrin V S, Kavokin A V 2010 *Superlattices and Microstructures* **47** 4449
[16] Angelis F, Das G, Candeloro P, Patrini M, Galli M, Bek A, Lazzarino M, Maksymov I, Liberale C, Andreani L, Di Fabrizio E 2010 *Nature Nanotechnology* **5** 67-72
[17] Gazizov A R, Zohrabi M, Kharintsev S S, Salakhov M Kh 2016 *J. Phys.: Conf. Ser.* **714** 012010
[18] Yablonovitch E 1987 *Phys. Rev. Lett.* **58** 2059-2062
[19] Ding B, Pemble M, Korovin V, Peschel U and Romanov S 2010 *Phys. Rev.* **82** 035119
[20] Lopez-Garcia M, Galisteo-Lopez J F, Blanco A, Sanchez-Marcos J, Lopez C and Garcia-Martin A 2010 *Small* **6** (16) 1757-1761
[21] Lopez-Garcia M, Galisteo-Lopez J F, Blanco A, Lopez C and Garcia-Martin A 2010 *Adv. Funct. Mater.* **20** 4338-4343
[22] Ding B, Bardosova M, Pemble M E, Korovin A V, Peschel U and Romanov S G 2011 *Adv. Funct. Mater.* **21** 4182-4192
[23] Romanov S G, Korovin A V, Regensburger A, Peschel U 2011 *Adv. Mater.***23** 2515-2533
[24] Galisteo-Lopez J F, Lopez-Garcia M, Lopez C and Garcia-Martin A 2011 *Appl. Phys. Lett.* **99** 083302-1-3
[25] Chen Y, Zhang D, Zhu L, Wang R, Wang P, Ming H, Badugu R and Lakowicz J R 2014 *Optica* **1** (6) 407-413
[26] Liu T-l, Russel K J, Cui S and Hu E L 2014 *Optics Express* **22** (7) 8219-8225
[27] Lin T, Lin J, Guo J, Kan H 2015 *Adv. Optical Mater.* **3** 1470-1475
[28] Frederich H, Wen F, Laverdant J, Coolen L, Schwob C and Maitre A 2011 *Optics Express* **19** (24) 24424-24433
[29] Afinogenov B I, Bessonov V O, Nikulin A A and Fedyanin A A 2013 *Appl. Phys. Lett.* **103** 061112

Light scattering in albumin solution under the influence of electric field in the regime of total internal reflection

E A Savchenko[1], E N Velichko[1], E T Aksenov[1], S V Rozov[1]
[1]Higher school of applied physics and space technologies, Institute of Physics, Nanotechnology and Telecommunications, Peter the Great Saint-Petersburg Polytechnic University, Saint-Petersburg, 195251, Russia

Abstract. In the paper the application of light scattering from biological macromolecules under the influence of electric fields in the regime of total internal reflection is discussed. An experimental setup on prism for the regime of total internal reflection was developed. The water solution of albumin with the concentration 4 % and pH of the medium is equal to 9.3 was used as an object of study. The obtained results demonstrate the applicability of method for study of biological macromolecules by light scattering in various fields of science.

1. Introduction

Biological macromolecules such as proteins, nucleic acids and polysaccharides have sizes from 1 nm to 1 μm. Particles with this sizes in liquids may be considered as colloidal systems and they carry a certain electrical charge due to electrical dissociation [1]. Each protein molecule contains many negatively and positively charged groups. Aggregation process, the environment and the pH of the medium have a great influence on charge of molecule. The structure and the properties of biological macromolecules change under the influence of various external factors.

Optical methods, including light scattering, are widely used to control these processes. Light scattering is the most common technique used to measure the parameters of biological macromolecules in solution [2]. Analysis of laser light scattering under the influence of electric fields allows us to measure both the electrophoretic mobility and diffusion coefficients of charged biological macromolecules in solution [3].

The goal of our work was to study electrophoretic mobility of biological macromolecules by means of light scattering under the influence of the electric fields. We considered the theory of light propagation in media and light scattering from particles in the regime of total internal reflection. The research possibilities of the developed experimental setup in analysis of biological samples in the regime of total internal reflection are considered. Such characteristic as electrophoretic mobility of molecules under the influence of various values of electric field strength for albumin solution were studied.

2. Techniques

In the electric field a double electric layer appears on the surface of particles (on the particle-liquid interface) [4]. A layer of ions of a certain sign uniformly distributed over the surface and creates a surface charge (potential-determining ions) on it. Ions of the opposite signs (counter ions) are attracted from the liquid medium to this layer. Motion of the particles leads to destruction of double electric field in some area (called the slip plane). The slip plane lies on the boundary between the diffuse and adsorption layers or in the diffuse layer near this boundary. The potential on the slip plane is called the electrokinetic or zeta potential [4,5]. The zeta potential corresponds to the stability of the biological macromolecules to aggregation. Molecules in electric fields move with the certain speed named electrophoretic mobility. Electrophoretic mobility relates to zeta potential.

Electrophoretic mobility of the solution can be measured by the method of light scattering. The intensity of the scattered light is detected using an optical system containing a photodetector and a signal processing system that allows spectral or correlation analysis of the data [2].

In the case of a field application, the particles begin to move at a certain speed to the oppositely charged electrode. The influence of the electric field is taken into account in the autocorrelation function [6]:

$$|g^{(1)}(\tau)| = \int_0^{\infty} F(\Gamma)e^{-\Gamma\tau} e^{iq\mu E \cos\frac{\theta}{2}} d\Gamma , \quad (1)$$

where Γ is the diffuse spectrum broadening, F(Γ) is the function contribution from the particles of one size into the total intensity of scattering, E is a value of electric field, μ is an electrophoretic mobility. The expression (1) can be used for calculation in case when the reorientation of particles and chemical reactions are not considered. In an electric field, the correlation function will be modulated by a cosine function, the frequency of which is determined by the electrophoretic mobility, and the damping rate is determined by the diffusion coefficient.

3. Experimental setup

The scheme of experimental setup for measurement of electrophoretic mobility is presented in fig.1.

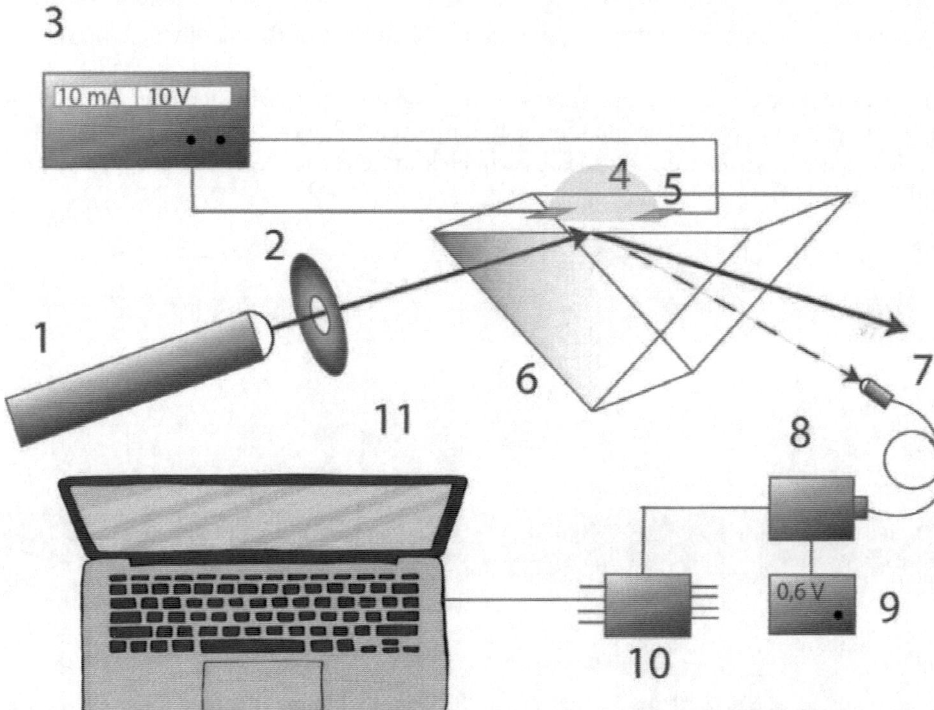

Figure 1. Experimental setup: 1 — laser, 2 — diaphragms, 3 — power supply, 4 —sample, 5 — electrodes, 6 — total internal reflection prism, 7 — optical fiber, 8 — photomultiplier, 9 — power supply of photomultiplier, 10 — ACD converter, 11 — computer

The area of investigation is located on prism (6). Two electrodes with sizes of 2*4 mm are placed on the top surface of the prism. The distance between the electrodes is 4mm. A sample is located on the gap between the electrodes. The volume of the sample is 10 μl.

IOP Conf. Series: Journal of Physics: Conf. Series **1068** (2018) 012011 doi:10.1088/1742-6596/1068/1/012011

A focused light from a semiconductor laser with wavelength λ = 655 nm is directed on the side face of the total internal reflection prism at the angles greater than the critical angle. The value of the critical angle is determined by the formula (2)

$$\theta_c = cos^{-1}\frac{n_2}{n_1}, \tag{2}$$

where n_1, n_2 are the refractive indices of the medium. In this case, the value of the critical angle was θ_c = 28.2 °.

The laser radiation incidents at angles greater than the critical angle. In this case radiation is reflected from the interface and is passed to the medium n_1. In optically medium with the smaller refractive indices n_2 value of electric field is limited. The penetration depth of radiation into the medium n_2 is defined as the distance at which the amplitude of the electromagnetic field in the optically less dense medium decreases in "e" times and depends on the incidence angle and the wavelength of the incident light. It is possible to change the incidence angle to study parameters of macromolecules in different depth. Such possibility increases the sensitivity and allows us to study the interaction of the sample with the surface.

The penetrated light is scattered from the sample and thus a speckle field is formed. The intensity is detected using a photoelectric multiplier placed at an angle of 5 degrees to the reflected light beam. Then the signal is passed to the ADC converter and to the computer for further processing the data.

4. Results

The water solution of albumin with the concentration 4 % and pH of the medium is equal to 9.3 was used as an object of study.

In the work we calculated autocorrelation functions for signals registered when different electric field were applied. Experimental results on albumin solution without external electric fields and in electric field E = 20 V/cm is shown in fig.2. In the experiments the electric field was changed in the range from 0 to 60 V/cm.

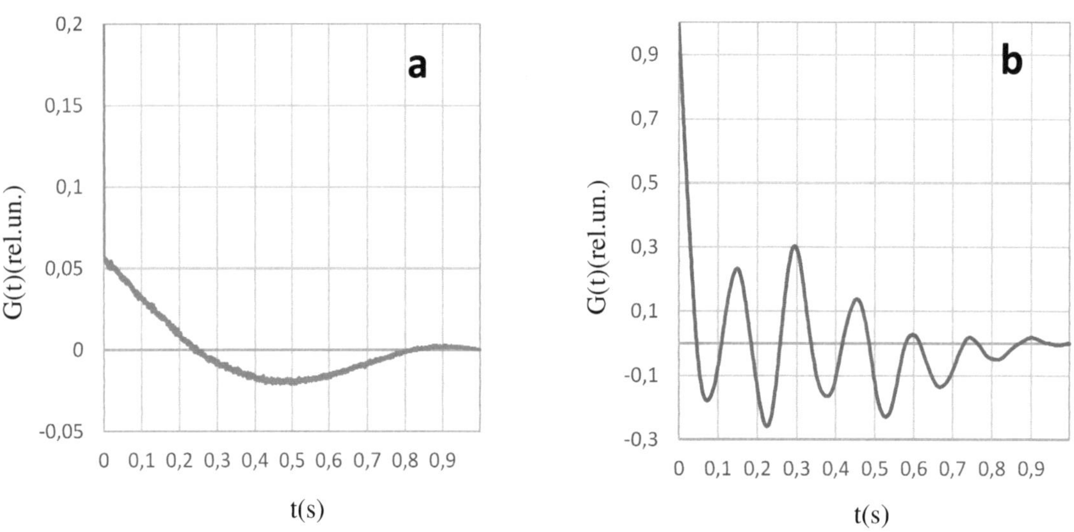

Figure 2. Autocorrelation functions of albumin solution in electric field: a) 0V/cm; b) 20 V/cm

The electrophoretic mobility in the case of a monodisperse solution can be calculated according to the relation (3)

$$\Delta t = \frac{2\pi}{\mu EK\cos\varphi/2}, \tag{3}$$

where Δt is the oscillation period of the autocorrelation function, the value of which can be found from the obtained experimental dependences, K is scattering vector, φ is the angle at which scattered radiation is recorded. Dependence of electrophoretic mobility on the amplitude of the electric field of albumin solution in pH = 9.3 is shown in fig.3.

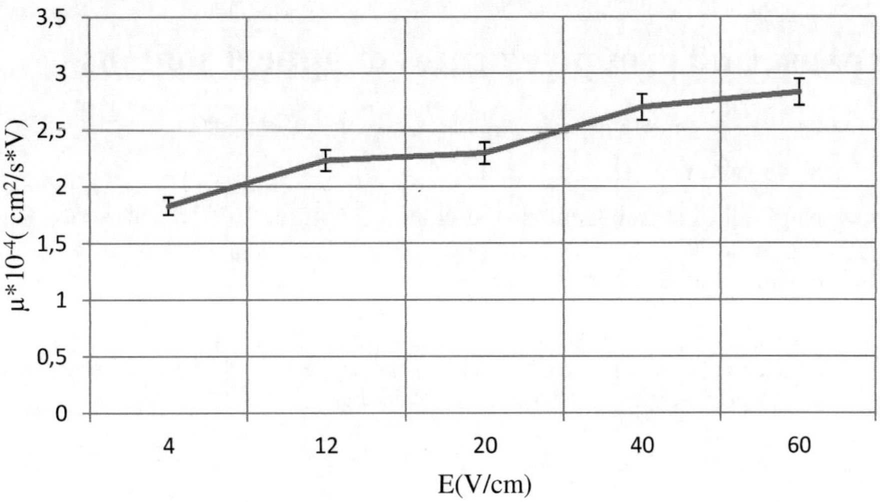

Figure 3. Dependence of electrophoretic mobility of albumin solution on the electric field strength

The electrophoretic mobility of the particles in the solution increases. The obtained results are in good correspondence with the theoretical data.

5. Conclusions

The results of experimental studies allow us to make a number of conclusions. Firstly, light scattering in the regime of total internal reflection makes it possible to study the dynamic characteristics of biological macromolecules in solution such as the electrophoretic mobility and diffusion coefficients. Secondly, the obtained autocorrelation functions allow us to calculate the electrophoretic mobility. An important advantage of the proposed method based on the regime of total internal reflection is possibility to analyze the parameters of small amount of the sample in express regime. In the measurement small values of electric field strength are used that has critical importance for the experiments with biological macromolecules.

References

[1] Bergethon P R 2013 *The physical basis of biochemistry: the foundations of molecular biophysics* (Springer Science & Business Media) p 566

[2] Nepomnyashchaya E K, Savchenko E A, Velichko E N, Aksenov E T 2017 *J. of Biomed. Phot. and eng.* **2** 040309

[3] Sourav B 2016 *J. Controlled Release* **235** 337

[4] Delgado A V, González-Caballero F, Hunter R J, Koopal L K, Lyklema J 2007 *J.of Colloid and Interf. Science* **309** 194

[5] Tereshchenko S A, Shalaev P V, Masloboev Yu P, Dolgushin S A, Deshabo V A, Yudin I K 2017 *J. Biomedical Engineering* **50** 333

[6] Serdyuk I N, Zaccai N R, Zaccai J 2017 *Methods in Molecular Biophysics: Structure, Dynamics, Function* (Cambridge University Press) p 702

IOP Conf. Series: Journal of Physics: Conf. Series **1068** (2018) 012012 doi:10.1088/1742-6596/1068/1/012012

On the phase and group velocities of optical solitons

S V Sazonov

National Research Centre «Kurchatov Institute», 123182 Moscow, Russia

e-mail: sazonov.sergey@gmail.com

Abstract. The method of analytical continuation of the dispersion parameters on the complex plane is analyzed. On this basis the new general formulae for phase and group velocities of the soliton-like pulses are obtained.

1. Introduction

One of the trends of modern nonlinear optics is the creation in the laboratory conditions of pulses of shorter durations τ_p. In the sixties of the last century, after the appearance of lasers with Q-switching and mode synchronization, pulses of nanosecond duration were generated. At the turn of the 70's and 80's in many laboratories in the world was overcome by a picosecond barrier. By now, we can confidently talk about femtosecond and even attosecond pulses [1]. The characteristic frequency for red color is $\omega \approx 3 \cdot 10^{15} s^{-1}$, which corresponds to the period $T \approx 2 \cdot 10^{-15} s$. The number of electromagnetic oscillations contained within the pulse can be estimated as $N \approx \tau_p / T \sim \omega\tau_p$. Then nano- and picosecond pulses contain, respectively, about one million and thousands of periods of light oscillations. Thus, here we can determine a small parameter $\mu \sim 1/N \sim (\omega\tau_p)^{-1}$. The presence of this small parameter makes it possible to use the approximation of a slowly varying envelope (SVE) [2].

A pulse lasting several femtoseconds contains about one period of electromagnetic oscillations. There is already a parameter μ of the order of unity and it is impossible to introduce the concept of an envelope. In this case, it is necessary to derive equations not for the envelope, but for the electric field itself of the light pulse [3 - 5].

With the shortening of the pulse duration with an unchanged period of electromagnetic oscillations, its spectral width increases. To a few-cycle pulse (FCP), the notion of the carrier frequency is not applicable. Therefore, well-known expressions for the phase and group velocities of a quasimonochromatic signal are also subject to revision. The generalization of these expressions to the case of short pulses is the goal of the work. To achieve this, a method of analytical continuation of dispersion parameters (ACDP) onto the complex plane will be used below.

2. Analytical continuation of dispersion parameters

The method presented in this section makes it possible to determine the dependence of the phase and group velocities of soliton-like pulses from their duration and the central frequency of the spectrum from the linear dispersion relation. It is sufficient to know that the investigated nonlinear equation has solutions in the form of localized traveling waves. As a rule, this localization has an exponential character, i.e. on the "tails" of the pulse, its field falls off as follows (assuming that the pulse propagates along the z-axis)

$$E \sim \exp\left[-\frac{\left|t - z / v_g\right|}{\tau_p}\right]\cos\left[\omega\left(t - z / v_{ph}\right)\right] = \mathrm{Re}\left\{\exp\left[i(\omega + i\gamma)t - (k + iq)z\right]\right\}, \quad (1)$$

where $\gamma = 1/\tau_p$, phase v_{ph} and group v_g velocities are determining by the expressions [6]

$$\frac{1}{v_g} = \frac{q}{\gamma}, \qquad \frac{1}{v_{ph}} = \frac{k}{\omega}. \quad (2)$$

Let us describe the ACDP procedure. On the "tails" of a localized pulse, the field is weak. Therefore, the linear approximation is applicable here. Let on the basis of the linearized version of the nonlinear equation under investigation, the following dispersion equation is derived:

$$k = k(\omega) \quad (3)$$

Next, the operation of the ACDP is carried out:

$$\omega \to \omega + i\gamma, \qquad k \to k + iq. \quad (4)$$

Substituting (4) into (3), we obtain $k + iq = k(\omega + i\gamma)$. Separating the real and imaginary parts, we arrive at two equations

$$k = k(\omega, \gamma), \qquad q = q(\omega, \gamma). \quad (5)$$

Substitution of these expressions into (2) leads to the dependences $v_g(\omega, \tau_p)$ and $v_{ph}(\omega, \tau_p)$.

In the case of quasimonochromatic pulses, the expression for the group velocity is well known. A natural question arises as to how this expression will change as the pulse duration becomes shorter. In the case of FCP, the notion of the carrier frequency is missing, but the notion of the center frequency of the spectrum remains. It is clear that the group velocity and phase velocity can no longer be calculated from the formulae $v_g = \partial\omega / \partial k$ and $v_{ph} = \omega / k$, respectively. The ACDP method is able to give an answer to this question. Let us note that $k = \mathrm{Re}\left[k(\omega + i\gamma)\right] = \left[k(\omega + i\gamma) + k(\omega - i\gamma)\right]/2$, $q = \mathrm{Im}\left[k(\omega + i\gamma)\right] = \left[k(\omega + i\gamma) - k(\omega - i\gamma)\right]/2i$. Then the last two formulae (2) are written in the form

$$\frac{1}{v_{ph}} = \frac{k(\omega + i/\tau_p) + k(\omega - i/\tau_p)}{2\omega}, \qquad \frac{1}{v_g} = \tau_p \frac{k(\omega + i/\tau_p) - k(\omega - i/\tau_p)}{2i}. \quad (6)$$

IOP Conf. Series: Journal of Physics: Conf. Series **1068** (2018) 012012 doi:10.1088/1742-6596/1068/1/012012

Introducing the translation operator in the frequency space by the formula $k(\omega \pm i/\tau_p) = \exp\left(\pm i\frac{1}{\tau_p}\frac{\partial}{\partial\omega}\right)k(\omega)$, we rewrite (6) in the form

$$\frac{1}{v_g} = \tau_p \sin\left(\frac{1}{\tau_p}\frac{\partial}{\partial\omega}\right)k, \qquad \frac{1}{v_{ph}} = \frac{1}{\omega}\cos\left(\frac{1}{\tau_p}\frac{\partial}{\partial\omega}\right)k. \tag{7}$$

These operator expressions generalize well-known formulae for the phase and group velocities of quasimonochromatic pulses. Here the parameter ω has the meaning of the central frequency of the pulse spectrum. In the limit $\omega\tau_p >> 1$, this parameter goes into the carrier frequency of a quasi-monochromatic signal.

The operator expressions (7) have the meaning of the corresponding Taylor series expansions:

$$\frac{1}{v_g} = \frac{\partial k}{\partial\omega} - \frac{1}{3!\tau_p^2}\frac{\partial^3 k}{\partial\omega^3} + \frac{1}{5!\tau_p^4}\frac{\partial^5 k}{\partial\omega^5} - \dots, \tag{8}$$

$$\frac{1}{v_{ph}} = \frac{k}{\omega} - \frac{1}{2!\omega\tau_p^2}\frac{\partial^2 k}{\partial\omega^2} + \frac{1}{4!\omega\tau_p^4}\frac{\partial^4 k}{\partial\omega^4} - \dots \tag{9}$$

Let us illustrate the application of the expressions obtained here to the example of solitons of some equations. The envelope of a nonresonant quasimonochromatic pulse in an isotropic medium with a weak nonlinearity and dispersion is described by a nonlinear Schrödinger equation

$$i\frac{\partial\psi}{\partial z} = -\frac{k_2}{2}\frac{\partial^2\psi}{\partial\tau^2} + \alpha|\psi|^2\psi, \tag{10}$$

where $k_2 = \partial^2 k/\partial\omega^2$ is the parameter of dispersion of the group velocity (DGV), α is the parameter of the Kerr nonlinearity, $\tau = t - z/v_{gl}$, v_{gl} is the linear group velocity determined by the first term in expansions of the right-hand side of expression (8).

Soliton solution of equation (10) looks like

$$\psi = \frac{1}{\tau_p}\sqrt{-\frac{k_2}{\alpha}}\exp\left(i\frac{k_2 z}{2\tau_p^2}\right)\operatorname{sech}\left(\frac{t - z/v_{gl}}{\tau_p}\right). \tag{11}$$

The phase factor in (11) corresponds to a nonlinear addition to the inverse phase velocity (or to the refractive index). In fact, taking (11) into account, we represent the phase φ of the light pulse in the form $\varphi = \omega t - kz + k_2 z/(2\tau_p^2) = \omega\left(t - z/v_{ph}\right)$. Hence we have

$$\frac{1}{v_{ph}} = \frac{k}{\omega} - \frac{k_2}{2\omega\tau_p^2}. \tag{12}$$

It is easy to see that expression (12) corresponds to the first two terms of the expansion on the right-hand side of expression (9). The group velocity of a soliton is equal to the linear group velocity and is determined only by the first term of the expansion on the right-hand side of expression (8).

To obtain the dependence of the group velocity on the duration, one can take into account the terms of a higher order in the equation for the envelope. This means that we must generalize equation (10). Taking into account the dispersion of nonlinearity and the linear dispersion of the group velocity of the third order, we obtain the Hirota equation [7]

$$i\frac{\partial\psi}{\partial z} = -\frac{k_2}{2}\frac{\partial^2\psi}{\partial\tau^2} + i\frac{k_3}{6}\frac{\partial^3\psi}{\partial\tau^3} + \alpha|\psi|^2\psi + i\beta|\psi|^2\frac{\partial\psi}{\partial\tau}, \tag{13}$$

where $k_3 = \partial^3 k/\partial\omega^3$, β is the coefficient characterizing the dispersion of the Kerr nonlinearity.

Under the condition $k_2\beta = -k_3\alpha$, equation (13) has a soliton solution of the form (11). In this case, the group velocity is given by the expression

$$\frac{1}{v} = \frac{1}{v_g} - \frac{k_3}{6\tau_p^2}. \tag{14}$$

It is easy to see that the right-hand side of this expression corresponds to the first two terms of the expansion on the right-hand side of (8). In this case, the phase velocity is still determined by the expression (12).

The procedure for further complicating the equation for the envelope can be continued. Then the terms of higher order with respect to the parameter μ will be consistently taken into account. A shorter and more rational procedure is that it is necessary to abandon the approximation of a slowly varying envelope. I.e., it is necessary to derive an equation for the electric field E of the pulse. In the case of a nonresonant nonlinear interaction with a dielectric, the dynamics of the pulse field is described by the equation [4]

$$\frac{\partial E}{\partial z} + aE^2\frac{\partial E}{\partial\tau} - b\frac{\partial^3 E}{\partial\tau^3} + g\int_{-\infty}^{\tau} E d\tau' = 0. \tag{15}$$

Here a is a coefficient that takes into account the Kerr nonlinearity, parameters b and g characterize the dispersion of the electron and ion responses, respectively, $\tau = t - n_0 z/c$, c is the speed of light, n_0 is the dispersionless refractive index.

Equation (15) does not have exact analytic solutions in the form of soliton-like pulses. However, in this case we can also find the dependences of the group and phase velocities on the central frequency ω of the pulse spectrum and its duration τ_p. The linearized version of equation (15) corresponds to the dispersion relation

$$k = \frac{n_0\omega}{c} + b\omega^3 - \frac{g}{\omega}. \tag{16}$$

Substituting this expression into (8) and (9) we find

$$\frac{1}{v_g} = \frac{n_0}{c} + b\left(3\omega^2 - \frac{1}{\tau_p^2}\right) + \frac{g}{\omega^2}\left[1 - \left(\frac{1}{\omega\tau_p}\right)^2 + \left(\frac{1}{\omega\tau_p}\right)^4 - \ldots\right],$$

$$\frac{1}{v_{ph}} = \frac{n_0}{c} + b\left(\omega^2 - \frac{3}{\tau_p^2}\right) - \frac{g}{\omega^2}\left[1 - \left(\frac{1}{\omega\tau_p}\right)^2 + \left(\frac{1}{\omega\tau_p}\right)^4 - \ldots\right].$$

Summing up here infinite geometric progressions, we have [8]

$$\frac{1}{v_g} = \frac{n_0}{c} + b\left(3\omega^2 - \tau_p^{-2}\right) + \frac{g}{\omega^2 + \tau_p^{-2}}, \qquad \frac{1}{v_{ph}} = \frac{n_0}{c} + b\left(\omega^2 - 3\tau_p^{-2}\right) - \frac{g}{\omega^2 + \tau_p^{-2}}. \quad (17)$$

Setting here first, and then, we arrive at the known expressions for the breather solutions of the modified Korteweg – de Vries equation [9] and the Schäfer – Wayne equation [10, 11], respectively.

The above geometric progressions converge, if $\omega\tau_p > 1$. On the other hand, when (16) is substituted into expressions (6), we arrive at the expressions (17) without using this restriction. Hence we come to the conclusion that formulae (6) are more general than (7) - (9). However, expressions (7), (8) and (9) are more intuitive, since it allows us to compare the general expressions for the group and phase velocities of solitons with the corresponding linear expressions. Here the situation is analogous to the hypothesis of "elimination of unpleasant divergences" in quantum field theory. This hypothesis allows us to summarize the Feynman diagrams of all orders in describing the systems of interacting quasiparticles [12].

3. Concluding remarks

Thus, in this paper a detailed analysis of the ACDP method is carried out. It is not always possible to find analytical solutions to these nonlinear wave equations or systems. It is in these cases that this method comes to the rescue.

The expressions (6) - (9) for the group and phase velocities can be used for both envelope pulses and FCP. In the first case, the terms containing the pulse duration should be considered as small corrections to the first terms on the right-hand sides of expansions (8) and (9) containing the carrier frequency. In the second case, the parameter has the meaning of the central frequency of the pulse spectrum and all terms of the series (8), (9) can make a significant contribution.

The procedure used in this article allows a quasiparticle interpretation of solitons. When $\gamma = q = 0$, we have a superposition of quasiparticles characterized by a linear dispersion equation. Accounting for nonlinearity leads to interaction between quasiparticles. As a result, the wave field is localized. The imaginary additions to ω and k in (4) should be treated as the inverse lifetime and the inverse mean free path of the quasiparticle. Thus, the duration of the soliton $\tau_p = 1/\gamma$ denotes the mean lifetime of quasiparticles in a state with energy $\hbar\omega$ and momentum $\hbar k$ ($\hbar$ is the Planck constant). When $\gamma/\omega << 1$, the quasiparticles interact weakly with one another, we have which corresponds to the envelope soliton. If $\gamma/\omega \sim 1$, we have a few-cycle pulse, consisting of quasiparticles with strong interaction.

Acknowledgment
This work is supported by the Russian Science Foundation (grant No. 17-11-01157).

References

[1] Kryukov P G 2008 *Femtosecond pulses* (Moscow: Fizmatlit)
[2] Kivshar Yu S and Agrawal G P 2003 *Optical solitons* (New York: Academic Press)
[3] Belenov E M, Nazarkin A V , and Ushchapovskii V A 1991 *Sov Phys JETP* **73** 422
[4] Kozlov S A and Sazonov S V 1997 *JETP* **84** 221
[5] Agrawal G P 1989 *Nonlinear fiber optics* (New York: Academic Press)
[6] Leblond H and Mihalache D 2013 *Phys Reports* **523** 61
[7] Sazonov S V 2018 Romanian Reports in Physics **70** 401
[8] Sazonov S V and Khalyapin V A 2003 *Optics and Spectroscopy* **95** 401
[9] Lamb G L 1980 *Elements of Soliton Theory* (New York: A Wiley)
[10] Schäfer T and Wayne C E 2004 *Physica D* **196** 90
[11] Sakovich A and Sakovich S 2006 *J Phys A: Math & Gen* **39** L361
[12] Mattuck R D 1967 *A guide to Feynman diagrams in the many-body problem* (London: McGraw Hill Publishing Company Limited).

Luminescence kinetics of CdSe-based quantum dots in toluene

A G Shmelev, D K Zharkov, A V Leontyev, V G Nikiforov, V S Lobkov and V V Samartsev

Kazan E. K. Zavoisky Physical-Technical Institute (KPhTI), Sibirsky tract, 10/7, 420029, Kazan, Russia

E-mail: sgartjom@gmail.com

Abstract. Luminescence kinetics of quantum dots (CdSe-CdS core-shell nanoparticles) in toluene were studied. Quantum dots were obtained through colloidal synthesis method in aqueous-organic media with the average size estimated as 2 nm and 2.9 nm from the position of the exciton absorption peak. A biexponential luminescence decays with time constants of a few and tens of picoseconds were observed via up-conversion technique.

1. Introduction

CdSe-CdS nanoparticles have gained a particular interest due to their unique optical properties [1]. The evolution of photoinduced charge carriers and, therefore, the luminescence kinetics strongly depends on the size of the nanocrystallite and the properties of its surface. As a rule, the reported optical responses decay within picoseconds which is associated with exciton-phonon interaction processes [2]. Auger recombination processes [3] can also occur, leading to rapid subpicosecond decay. Generally, multicomponent luminescence decays are being observed, e.g. [1], with time constants up to hundreds of nanoseconds [1], depending on the quantum dot sizes.

The problem of the quantum dots synthesis is the aggregation of single dots into clusters of tens and hundreds of nanometers. This process greatly suppresses the confinement effect, and therefore, makes it less appealing to applications in optoelectronics. To address this problem, we have obtained quantum dots of 2 nm and 2.9 nm in size, consisting of CdSe cores coated with CdS shell. Those are less likely to get aggregated, while retaining all the confinement-related optical properties.

2. Experimental

CdSe nanoparticles were synthesized using colloidal method in aqueous organic media [4]. Cadmium acetate and sodium selenosulfate were used as a source of cadmium and selenium ions. In order to obtain quantum dots (QD) dispersible in organic solvents, like toluene, hexane, chloroform, etc., oleic acid was chosen as the stabilizer. The organic component of the medium necessary to dissolve the oleic acid was ethanol or glycerin. Depending on the required optical characteristics, the synthesis of CdSe QD was carried out at different temperatures.

CdSe nanocrystallites with a luminescence maximum at 500 nm (Fig.1 (3)) were obtained in a water-ethanol medium at 50°C. The QD were purified by reprecipitation from a mixture of ethanol-hexane. The absorption spectrum of QD contains a wide wing in the range of 400-500 nm, corresponding to exciton absorption. The average size of QD was estimated from the position of the exciton absorption peak at 457 nm [5]. According to these values, the average size of CdSe QD was 2 nm.

IOP Conf. Series: Journal of Physics: Conf. Series **1068** (2018) 012013 doi:10.1088/1742-6596/1068/1/012013

CdSe nanocrystallites with a luminescence maximum at 570 nm were obtained in an aqueous glycerin medium at 170 ° C. These nanocrystallites absorb light in a wider range, and the exciton peak is at 543 nm (Fig. 1 (4)). The average size of CdSe QD was estimated as 2.9 nm.

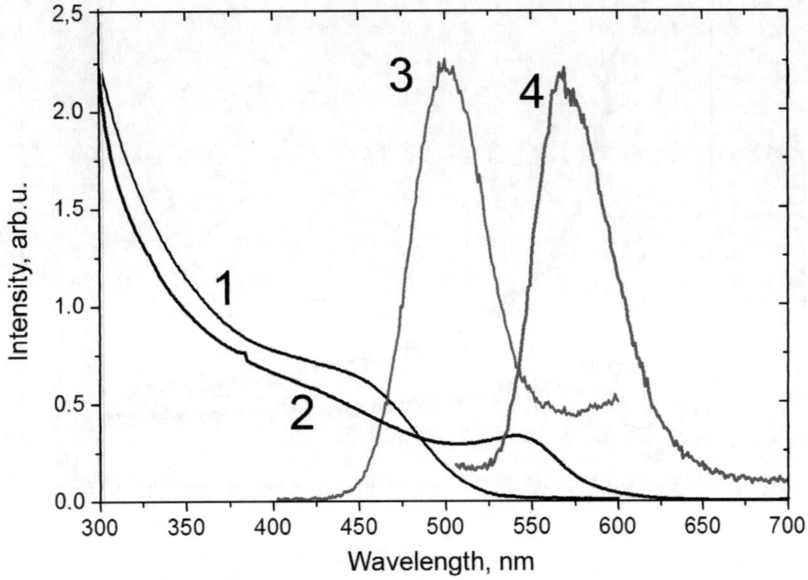

Figure 1. Absorbance (1,2) and luminescence (3,4) spectra of 2 nm (1,3) and 2.9 nm (2,4) CdSe-CdS QD in toluene.

Time-resolved fluorescence up-conversion technique [6] was applied to both samples. To excite the luminescence of the samples, the second harmonic of Ti:Sapphire laser (pulse duration of 50 fs, carrier wavelength of 770 nm, average power of 450 mW) was used with a wavelength of 385 nm and the average power of 20 mW. A suspension of QD in toluene was pumped through a quartz cuvette 2 mm thick. The remaining part of 770 nm laser beam with an average power of 330 mW was passed through the optical delay line and used as a temporal gate. The luminescence emission was collected by a parabolic mirror and, along with the gate beam, focused onto a BBO type 1 crystal of 0.5 mm thickness, where sum frequency signal was generated. This signal was isolated by a monochromator and recorded by the photon counting system.

3. Results

The recorded experimental data (Fig. 2) is the convolution of the QD luminescence emission and the gate pulse. The width of spread function can be estimated as 100 fs (considering the duration of femtosecond pulses), which is significantly less than typical luminescence decay times. It was established [2-3] that different relaxation processes of photoexcitation must predominate depending on the QD size, namely, electron-phonon relaxation for larger dimensions and Auger recombination for smaller crystallites. As we have studied QD of two sizes, 2 nm and 2.9 nm, we expected to find a significant difference in the ratio of "faster" (sub-picosecond) and "slower" (tens of picoseconds) contributions.

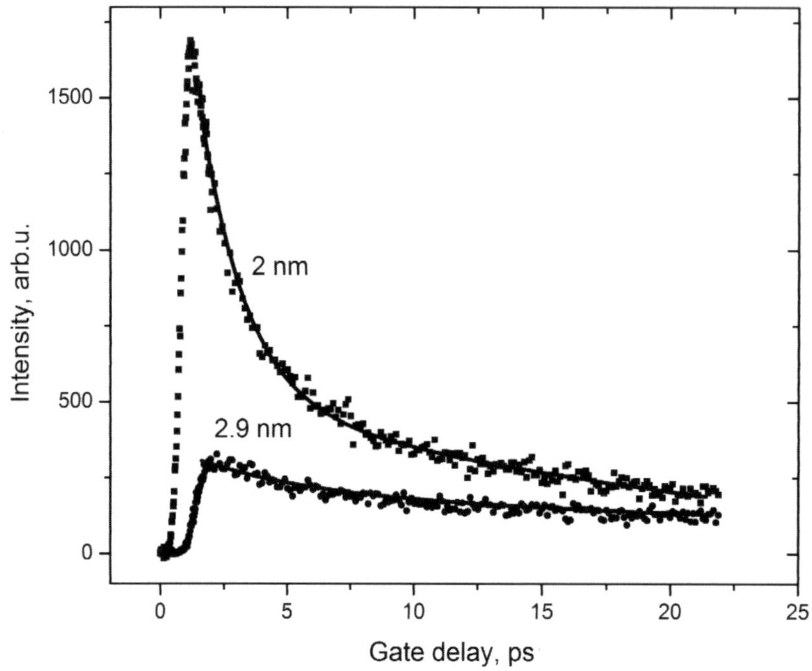

Figure 2. Luminescence kinetics of 2 nm and 2.9 nm CdSe-CdS QD in toluene.

As evident from Fig. 2, the basic shape of the luminescence kinetics curve does not depend on the size of the QD, so the relaxation processes for larger and smaller QD are arguably the same. However, the parameters of the two-exponential decay of the luminescence kinetics depend significantly on the size of the QD: the decay times of 1.8 ps and 26.8 ps correspond to 2nm QD, and 4.5 ps and 68 ps are detected for 2.9 nm QD. We should note that in a single CdS crystal the carrier thermalization time within the conduction band is of the same order [7], which may indicate some agglomeration of QD in the suspension.

4. Conclusion

The application of time-resolved fluorescence up-conversion technique revealed a two-component nature of the luminescence decay. For QD with average size estimated as 2 nm the decay times were found to be 1.8 and 26.8 ps, and for 2.9 nm QD the decay times were 4.5 and 68 ps. The established correlation between the decay times and the size of the QD fits with the existing data [2-3]. More detailed analysis of the dependence of the relaxation processes in QD on their size will require additional experimental and theoretical studies.

References

[1] Resch-Genger U, Grabolle M, Cavaliere-Jaricot S, Nitschke R, Nann N 2008 *Nature Methods* **5** 763
[2] Prabhu S S, Vengurlekar A S, Shah J 1996 *Physical Review B* **53** R10465
[3] Klimov V I , McBranch D W 1998 *Physical Review Letters* **80** 4028
[4] Shukla R K, Galyametdinov Yu G, Shamilov R R, Haase W 2014 *Liquid Crystals* **41** 1889
[5] Yu W W, Qu L, Guo W, Peng Y X 2003 *Chemistry of Materials* **15** 2854
[6] Kahlow M A, Jarzeba W, DuBruil T P, Barbara P F 1988 *Review of Scientific Instruments* **59** 1098
[7] Shmelev A G, Leontyev A V, Nikiforov V G, Ivanin K V, Lobkov V S, Khasanov O Kh, Samartsev V V 2015 *Journal of Physics: Conference Series* **613** 012014